SpringerBriefs in Applied Sciences and Technology

Display Science and Technology

Series editors

Karlheinz Blankenbach, FH für Gestaltung, Technik, Hochschule Pforzheim FH für Gestaltung, Technik, Pforzheim, Germany

Fang-Chen Luo, Hsinchu Science Park, AU Optronics Hsinchu Science Park, Hsinchu, Taiwan

Barry Blundell, St-Priest-la-Marche, France

Robert Earl Patterson, Human Analyst Augmentation Branch, Air Force Research Laboratory, Wright-Patterson AFB, OH, USA

Jin-Seong Park, Science and Engineering, Hanyang University, Division of Materials Science and Engineering, Seoul, Korea (Republic of)

W0235367

This series presents readers with concise, readable books describing advances and the state-of-the-art in the displays field. Featuring compact volumes of 75-125 pages, it forms a companion to the *Series in Display Science and Technology*, covering the same range of topics, from the fundamentals of optics, color science and human factors, through display materials, electronics and driving, to advances in display technologies including LCDs, OLEDs, reflective displays, 3D displays, mobile displays and more. Related fields such as display metrology, human-computer interaction and energy are also covered.

More information about this series at http://www.springer.com/series/15635

Katsuyuki Morii • Hirohiko Fukagawa

Air-Stable Inverted Organic Light-Emitting Diodes

 Springer

Katsuyuki Morii
Advanced Material Research Center
Nippon Shokubai Co. Ltd.
Suita, Japan

Hirohiko Fukagawa
Science & Technology Research
Laboratories
Japan Broadcasting Corporation
Setagaya-ku, Tokyo, Japan

ISSN 2511-1434   ISSN 2511-1442 (electronic)
Display Science and Technology
ISSN 2191-530X   ISSN 2191-5318 (electronic)
SpringerBriefs in Applied Sciences and Technology
ISBN 978-3-030-18513-8   ISBN 978-3-030-18514-5   (eBook)
https://doi.org/10.1007/978-3-030-18514-5

This Springer imprint is published by the registered company Springer Nature Switzerland AG
The registered company address is: Gewerbestrasse 11, 6330 Cham, Switzerland

Contents

1   **Air Stability for Organic Light-Emitting Diodes**   1
    1.1   Comparison With LEDs   1
    1.2   Important Issues for OLEDs   2
    1.3   Conclusion   3
    References   3

2   **Hybrid Organic-Inorganic Light-Emitting Diode**   5
    2.1   First Hybrid Organic-Inorganic Light-Emitting Diode (HOILED) . . .   5
    2.2   Metal Oxide Layer on the Cathode Side for HOILEDs   7
        2.2.1   Zinc Oxide   7
        2.2.2   Wide-Band-Gap Metal Oxides   9
    2.3   Initial Modification of Surface of Metal Oxide
        on the Cathode Side   10
    2.4   Conclusion   11
    References   12

3   **Interfacial Engineering by Introducing an Interlayer**   13
    3.1   Self-Assembled Monolayers   13
    3.2   Ionic Species   15
    3.3   Polyethylenimine Ethoxylated (PEIE) and Branched
        Polyethylenimine (PEI)   18
    3.4   Other Amine Derivatives   26
    3.5   Novel Metal Oxides and Electrodes   28
    3.6   Solution-Processed Boron Compounds for Operationally
        Stable Inverted OLEDs   29
    3.7   Conclusion   31
    References   31

4 Carrier Injection Mechanism 33
   4.1 Basic Carrier Injection Mechanism and Related Electronic
       Structures 34
   4.2 Electron Injection Mechanism in Inverted OLEDs 40
   4.3 Hole Injection Mechanism in Inverted OLEDs 42
       4.3.1 Organic/Inorganic Layer and Inorganic/Organic Layer...... 44
       4.3.2 MoO_3 on F8BT..................................... 44
   4.4 Conclusion 47
   References... 47

Chapter 1
Air Stability for Organic Light-Emitting Diodes

Abstract We introduce the importance of air stability on the basis of the principles and the history of organic light-emitting diodes (OLEDs), and the way of realising air-stable OLEDs finally. OLEDs are current-driven self-emitting devices that, in principle, have the features of lightness and thinness. Therefore OLEDs are expected to have unprecedented flexibility. However, it is difficult to achieve air stability in OLEDs, which is the key property for the realisation of flexible devices, because it has been essential to use air-active materials in consideration of the operational mechanism.

The trend toward lighter products, thinner devices, and greater toughness is inevitable. Lightweight products conserve resources, leading to an environmentally friendly society. Products based on thin devices are attractive to consumers, leading to a creative society. Tough devices increase the reliability of products, ensuring a safe society. In the field of electronic devices, substrates have shifted from metals to glass, and now to polymers, and we are now in a transition period between the use of glass and polymer substrates. Organic electronics have been developed during this period and their future development in line with the above trend is expected. Although organic light-emitting diodes (OLEDs) are one of the most promising types of device, there are major challenges to be overcome in their fabrication. The stability of the performance of OLEDs in air has been an important and major challenge since the principles of existing organic EL devices have been published. OLEDs are current-driven self-emitting devices and also serve as practical devices, in which carriers move through the organic/organic interface and through the organic/inorganic interface[1–4]

1.1 Comparison With LEDs

OLEDs have an advantage over inorganic light-emitting diodes (LEDs) in terms of the layer thickness. OLEDs realise carrier injection and transport, as well as the recombination of carriers, owing to their ultrathin layer structure and controlled interface. The layer thickness of an OLED is less than 100 nm and the thinnest layer is a monomolecular layer. In other words, multiple properties of OLEDs are affected

K. Morii, H. Fukagawa, *Air-Stable Inverted Organic Light-Emitting Diodes*,
SpringerBriefs in Applied Sciences and Technology,
https://doi.org/10.1007/978-3-030-18514-5_1

by the characteristics of molecules. One such property is carrier injection, which will be explained in detail in Chap. 3.

In principle, most OLEDs can be fabricated in a temperature range from room temperature to less than 200 °C, in which organic compounds are not decomposed. This is markedly different from the case of inorganic LEDs, which require a temperature of more than 200 °C, and typically over 500 °C, to induce crystallisation on a [5]. Therefore, the expected applications of OLEDs and LEDs also differ. For OLEDs in which polymer film substrates are used, various applications that cannot be realised using conventional inorganic devices are expected owing to their flexibility and thin-layer structure. However, there is a problem of encapsulation before these applications can be realised.

1.2   Important Issues for OLEDs

One of the important papers on organic EL devices was written by Dr. Tan [1]. In his paper, two important issues to be resolved for the future development of organic EL devices are explained. One is the reduction of thickness and the other is the reduction of the injection barrier. In his paper, the possibility of thin functional layers with a thickness of ≤ 100 nm, each having different functions, was demonstrated as an alternative to conventional organic devices consisting of a single functional layer with a thickness of 2–3 μm [3, 4]. Owing to this structure, present organic EL devices have a total thickness of only 200–300 nm, while realising required functions using several stacked layers. This suggests that even a hard material with a high Young's modulus can be bent [6] paving the way to a new field of flexible optical devices. The other issue, i.e., the reduction of the injection barrier, is also important. To reduce the injection barrier, an active carrier injection source with a high energy level, in other words, a low work function, is necessary as a cathode. In this way, the OLED can be flexible, and at the same time, an encapsulation structure that protects the active cathode against water and oxygen is indispensable. In fact, the main part of degradation due to water and oxygen is near the cathode [7]. For example, electron injection layers such as LiF are very sensitive to air and can be severe problem. The encapsulation structure is made of a material having low water vapour transmission rate (WVTR). Most materials with low WVTR are rigid materials such as metal and glass, and are incompatible with the flexibility of the device. In Dr. Tan's paper, Mg was used as the source. Thus, encapsulation to achieve the detection limit of the WVTR of 10^{-6} g/m^2/d is required for OLEDs [8, 9]. This is a serious problem that prevents the realisation of flexibility because glass is the only optically feasible material to have this property. Therefore, the development of barrier films has been actively carried out. Several organic and inorganic thin layers have been laminated to add flexibility and barrier properties, respectively, to realize an ultra-high barrier property with a WVTR of 10^{-6} [10, 11]. With this structure, water molecules are blocked by a maze effect. The lack of inorganic layers leads to deterioration of the barrier property. Therefore, it is difficult to obtain a large-area barrier film for large-area panels, such as for organic TVs. Even today, the basic concept of realizing a high barrier property remains the same as when it was first proposed. Note that Dr. Tan attempted to increase the stability of the performance of organic EL devices in air by mixing Ag as an air-stable material into Mg. Achieving a stable performance is a severe problem for displays [8, 9], which are the most suitable application of organic EL devices. Flexible organic EL devices are expected to be used in various applications, and prototypes at an early stage of development have been demonstrated; however, they have not yet been practically applied to commercial products.

Attempts to achieve the stable performance of OLEDs in air are closely related to the development of interface engineering. Ishii et al. [12] reported that the distribution of localised electrons, such as in dipoles and chemical bonds, which is a characteristic of organic molecules, specifically affects the organic/inorganic layer interface. The shift of the vacuum level, which is small in inorganic substances, is large in organic substances because of the localised electrons. Campbell et al. [13] have demonstrated the control of the work function of Ag by adsorbing polar alkanethiol derivatives with various magnitudes and directions of the dipole moment [from 2.24D for $CH_3(CH_2)_9SH$ to -1.69D for $CF_3(CF_2)_7(CH_2)_2SH$] to form self-assembled monolayers (SAMs). These derivatives induced a change in the work function of 1.55 eV (from -0.7 to 0.85 eV). These characteristics exist in principle in devices using organic and inorganic thin films, although they can be advantageous for high-performance devices requiring a reduced injection barrier.

1.3 Conclusion

In this article, OLED devices having an inverted structure are discussed as a means of realising ideal flexible OLEDs. The development of hybrid organic-inorganic light-emitting diodes (HOILEDs) as a trigger for the development of the inverted structure is explained in the next chapter. In particular, key breakthroughs in the initial development phase are introduced with some examples, focusing on charge injection using oxides and modification of the surface of oxides. In Chaps. 3 and 4, the materials and principles of the latest inverted organic light-emitting diodes (iOLEDs) are respectively explained. Finally, the current situation and future prospects are summarised.

References

1. Tang, C. W. et al. Organic electroluminescent diodes. *Appl. Phys. Lett.* **51**, 913 (1987).
2. Nicht, S. et al. Light-emitting diodes based on conjugated polymers. *Nature* **347**, 539–541 (1990).

3. Mitschke, U. et al. The electroluminescence of organic materials. *J. Mater. Chem.* **10**(7), 1471–1507 (2000).
4. Friend, R. H. et al. Electroluminescence in conjugated polymers. *Nature* **397**, 121–128 (1999).
5. Williams, E. W. Luminescence and the light emitting diode. *publisher: Robert Maxwell, M.G.* Chapter 3 (Crystal Growth) and Chapter 4 (Fabrication).
6. Fukuda, K. et al. Free-standing organic transistor and circuit with sub-micron thickness. *Sci. Rep.* **6**, 27450 (2016).
7. Aziz, H. et al. Degradation processes at the cathode/organic interface in organic light emitting devices with Mg:Agcathodes. *Appl. Phys. Lett.* **72**(11), 2642–2644 (1999).
8. Jarvis, K. L. et al. Growth of thin barrier films on flexible polymer substrates by atomic layer deposition. *Thin solid films* **624**, 111–135 (2017).
9. Wang, L. et al. Enhanced moisture barrier paerformance for ALD-encapsulated OLEDs by introducing an organic protective layer. *J. Mater. Chem. C* **5**, 4017–4024 (2017).
10. Mitzi, D. B. et al. Thin-films deposition of organic-inorganic hybrid materials. *Chem. Mater.* **13**, 3283–3239 (2001).
11. Chatham, H. Oxygene diffusion barrier properties of transparent oxide coatings on polymeric substrate. *Surf. Coat. Tech.* **78**, 1–9 (1996).
12. Ishii, H. et al. Energy level alignment and interfacial electronic structures at organic/metal and organic/organic interfaces. *Adv. Mater.* **11**(8), 605–625 (1999).
13. Campbell, I. H. et al. Controlling schottky energy barriers in organic electronic devices using self-assembled monolayers. *Phys. Rev. B* **54**, R14321 (1996).

Chapter 2
Hybrid Organic-Inorganic Light-Emitting Diode

Abstract To achieve air stability in organic light-emitting diodes (OLEDs), a metal oxide layer was introduced on the cathode side. In these novel OLEDs, referred to as hybrid organic-inorganic light-emitting diodes (HOILEDs), an inverted structure was better than the conventional one from the viewpoint of the fabrication process. These HOILEDs have insufficient performance for practical use, even though the electron-injection barrier was reduced by the insertion of the metal oxide layer. In this chapter, by introducing the history of development of HOILED, we will explain the purpose of inverted OLED and the essential problem as the cornerstone of current inverted OLED (iOLED).

2.1 First Hybrid Organic-Inorganic Light-Emitting Diode (HOILED)

The reason why organic light-emitting diodes (OLEDs) lack robustness against oxygen and water is that they have an active layer. The area exhibiting most degradation by air is around the cathode [1]. On the basis of ref. [1], we replaced the cathode with an air-stable electrode having a large work function, similarly to the anode. As a result, as shown in Fig. 2.1, a large injection barrier occurred around the cathode. To reduce the electron injection barrier, a metal oxide was introduced on the cathode side [2]. It has already been reported that metal oxides can be used for OLEDs, but improvements have only been made on the anode side [3] and not on the cathode side. On the other hand, the electron transport in a metal oxide in dye-sensitized solar cells (DSSCs) fabricated at room temperature in air has been realized, although it occurred in the opposite direction to that in OLEDs [4]. The introduction of air-stable cathodes will be possible if the problem of the high injection barrier can be solved. As one of the solutions, with reference to DSSCs, a configuration in which a compact layer of titanium oxide was prepared on a cathode and a porous layer of titanium oxide was prepared on the top of the compact layer was fabricated. However, because the metal oxide layer was thick, the resulting devices had a high threshold voltage and most of the devices were short-circuited owing to the rough surface. The reason why an OLED is one order of magnitude thinner than DSSCs. Then, we fabricated a device without the porous TiO_2 layer,

© The Author(s) 2020
K. Morii, H. Fukagawa, *Air-Stable Inverted Organic Light-Emitting Diodes*,
SpringerBriefs in Applied Sciences and Technology,
https://doi.org/10.1007/978-3-030-18514-5_2

Fig. 2.1 Energy diagram around cathode in OLEDs; Work function dependence of cathode of energy barrier in electron injection

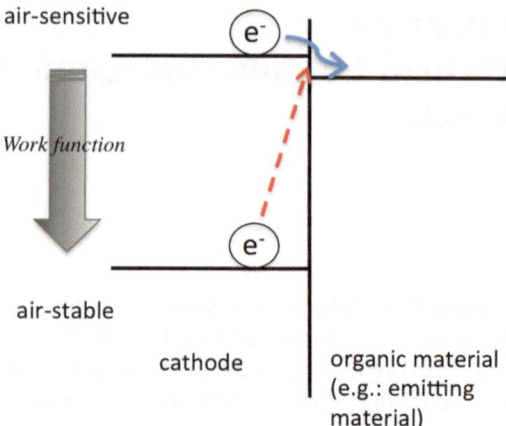

which only had the compact TiO_2 layer. This was the first hybrid organic-inorganic light-emitting diode (HOILED). For this device, higher brightness were sometimes observed in the second scan than in the first scan. This phenomenon was observed upon changing the fabrication process of the oxide layer on the cathode side. This suggests that a slight chemical difference on the surface of the metal oxide greatly affects the electron injection behaviour. Therefore improved EL characteristics were expected by modification of the metal oxide surface.

In the initial structure, a gold electrode as an anode was directly formed on poly(dioctylfluorene-alt-benzothiadizole) (F8BT), which is a light-emitting polymer, and the emission was observed at a voltage of only 10 V, although it was unstable. It was not possible to measure the J-V characteristics of the device. As an improvement on the anode side, carrier injection using the metal oxides reported by Tokito et al. [3] and a low turn-on voltage were realised. Although we initially used vanadium oxide (V_2O_5), it was replaced with molybdenum oxide (MoO_3) to improve safety and stability [5]. During this development, we observed a device that emitted light with high brightness at an extremely low turn-on voltage. This was the first HOILED with a high electroluminescence (EL) performance.

As a result, an HOILED, which was the origin of the iOLEDs, was realized. It had an inverted structure. Also, according to its band diagram, the HOILED was a hole-dominated device at the time of its conception. Details are given in Chap. 4. The achievement of the HOILED led to unique hole injection resulting from both the inverted structure and the introduction of a metal oxide layer, leading to the problem of achieving smooth electron injection under a high injection barrier.

Details of the mechanism of hole injection will be given in Sect. 4.3, and the development of new materials to improve electron injection will be described in detail in Chaps. 3 and 4. Next, we will discuss the initial introduction of the metal oxide on the electron injection side [6]. We will also briefly touch on early attempts to modify oxides that later led to iOLEDs. The air stability of the early HOILEDs will also be discussed.

2.2   Metal Oxide Layer on the Cathode Side for HOILEDs

In 2006, Morii et al. reported an HOILED with a metal oxide layer as the electron injection layer (EIL) as an inverted OLED [2]. The most important concept of this device is the introduction of a metal oxide layer for electron injection, where the inverted structure was merely used as a means of achieving this concept. The model for this device is the DSSC. Therefore, TiO_2 was used as the metal oxide. The first device structure was $FTO/TiO_2/F8BT/MoO_3/Au$. An EL image of this device is shown in Fig. 2.2. TiO_2 prepared by spray pyrolysis using a 0.18 M di-isopropoxy titanium bis(acetylacetonate) solution in ethanol typically used in a DSSC was employed [7] on the cathode side as the first layer. On the other hand, MoO_3 was used on the anode side, and the green-emitting polymer F8BT was used as the emitting layer. The HOILED had a quite low threshold voltage of below 2 V with a high brightness of about 1000 cd/m^2 at 4 V, although it is low efficiency (Fig. 2.3). The conduction band for TiO_2 was estimated to be around 4 eV, and the lowest unoccupied molecular orbital (LUMO) level for the emissive materials was approximately 3 eV (Fig. 2.4). It is difficult to explain the electron injection of this HOILED using the flat-band model.

2.2.1   Zinc Oxide

An alternative metal oxide to TiO_2 for electron injection in HOILEDs is zinc oxide (ZnO) owing to its slightly higher energy level (4 eV) and superior optical properties. A higher brightness and lower turn-on voltage were observed in ZnO-based devices than in TiO_2-based devices. This is suggested to be due not only to the high transparency and conductivity but also to the higher polarity on the surface [8]. In addition, the turn-on voltage in the ZnO-based devices was low compared with the energy of the emission of F8BT (2.4 eV). ZnO has a large number of oxygen vacancies, which form a level in a band gap. In the field of LEDs, Auger-mediated energy

Fig. 2.2 EL image of the first HOILED

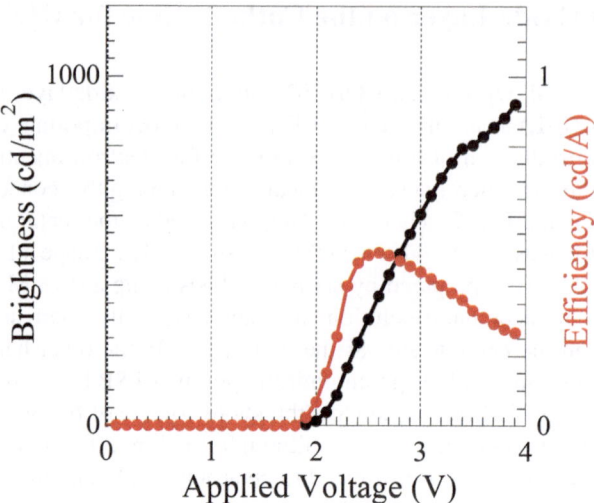

Fig. 2.3 LE–V characteristics of the first HOILED

Fig. 2.4 Band diagram
around cathode side of the
first HOILED

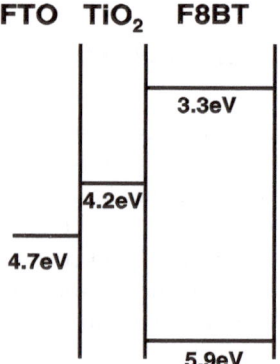

up-conversion using the level in a band gap has been shown, for example, in InP-
AlInAs [9]. Investigation of electro-injecting metal oxides in place of ZnO, for
example searching materials with multi-component metal oxides, was almost no
longer performed as the control of the oxide surface has become more important,
and the interest of research moved to focus on the development of organic materials
on oxides.

On the other hand, investigation on the correlation between the structure of metal
oxide film and device physical properties have been carried out separately by some
film fabrication processes. For an OLED using ZnO nanoparticles as an EIL [10], a
similar up-conversion mechanism as above was proposed. Thus, ZnO is promising
metal oxide layer for electron injection. However, the efficiency of inverted OLEDs

is still low compared with that of conventional OLEDs, even if a ZnO layer is used. One of the reasons for this is current leakage through the metal oxide. Therefore, metal oxides with a wide band gap were employed in an attempt to improve the device performance.

In addition, zinc oxide layers for OLEDs have been formed by various fabrication processes. The device performance depends on the fabrication process of the metal oxide layer. High conductivity and high brightness have also been reported for an inverted device [11], mainly due to the increase in the surface area. However, in some cases devices are unstable, depending on their structure.

2.2.2 Wide-Band-Gap Metal Oxides

Several different metal oxides fabricated by various processes (a solution process using a precursor or a dispersion of nanoparticles [10, 11], or gas phase deposition) have been used for the inorganic EIL. The behaviour of zirconium oxide (ZrO$_2$) as an electron-injecting material was reported by Tokmoldin et al. [12]. ZrO$_2$ is a typical wide-band-gap material with a high conduction band energy and a deep valence band energy that efficiently blocks holes on the metal oxide interface. Therefore, high efficiency has been achieved for ZrO$_2$-based devices with a maximum efficiency of approximately 3 cd/A around 10,000 cd/m^2 (Fig. 2.5). In analogy to the work on ZrO$_2$, as an alternative insulating metal oxide, the use of magnesium oxide (MgO) was evaluated, which has a higher LUMO and deeper highest occupied molecular orbital (HOMO) than ZrO$_2$ [12], because of its lower expected exciton-quenching effect and higher expected hole-blocking effect than those of ZrO$_2$. Efficiency of up to 3.5 cd/A at a high luminance was achieved. Hence, the use of insulating (wide-band-gap) metal oxides as the electron-injecting contact in HOILEDs is a promising alternative to the employment of more traditional materials such as ZnO and TiO$_2$. However, the duration of operational stability was also

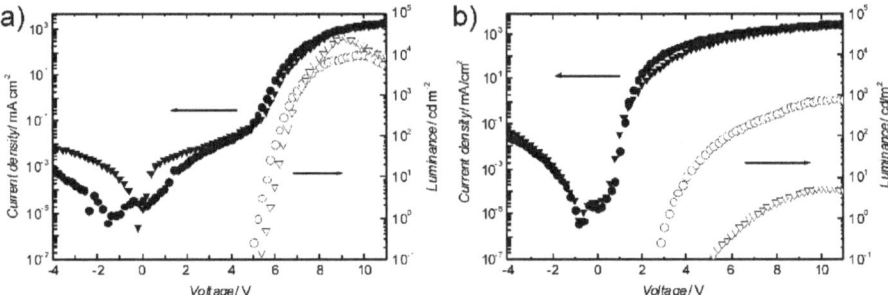

Fig. 2.5 (a) JL-V characteristics of ITO/ZrO2/F8BT/MoO3/Au with annealed (triangles) and unannealed (circles) polymers (b) JL-V characteristics of ITO/TiO2/F8BT/MoO3/Au with annealed (triangles) and unannealed (circles) polymers. Reproduced with permission from ref. [12]. Copyright 2009, Wiley-VCH

short in the present system because degradation of the emitting layer on the surface of the metal oxide could not be prevented.

2.3   Initial Modification of Surface of Metal Oxide on the Cathode Side

The operational mechanism in HOILEDs has also been studied. Holes in an HOILED were induced by MoO_3 [13]. As mentioned above, it was clarified that an HOILED is a hole-dominated device and that it is difficult to inject electrons into the device. Morii et al. then attempted to modify the surface of the metal oxide layer. The electron injection was improved by the use of Cs compounds obtained from Cs_2CO_3 [14]. The modified HOILED exhibited similar characteristics to a conventional OLED consisting of F8BT and poly(3,4-ethylenedioxythiophene):poly(styrene sulphonic acid) (PEDOT:PSS). Hsieh et al. reported improved electron injection in HOILEDs by using a dipole consisting of a self-assembled monolayer prepared on a metal oxide layer [15]. In the same way, reduction of the electron injection barrier have been reported by Lee et al. and Kim et al. The methods are the polar solvent treatment on ZnO [16], and insertion of polyethyleneimine layer into ZnO / Organic interface [17]. Further details are given in Chap. 3. Several trials based on the control of the surface energy have been reported. However, there have been few reports on the air-stability and operational stability of inverted OLEDs.

Although it is very primitive experiments, in parallel with the submission of the first efficient HOILEDs paper, the stability of the HOILEDs modified with Cs compounds in air has been evaluated. The JL-V characteristics in the initial state and after exposure to air for 172 h were reported. Unfortunately, some dark spots were observed and both the brightness and the current density degraded after 172 h in Fig. 2.6. However, the emission was still observable after 500 h in air. On the other

Fig. 2.6 Time dependence of JL-V characteristics of HOILEDs without compact TiO2 layers. Reproduced with permission from ref. [14]. Copyright 2008, American Institute of Physics

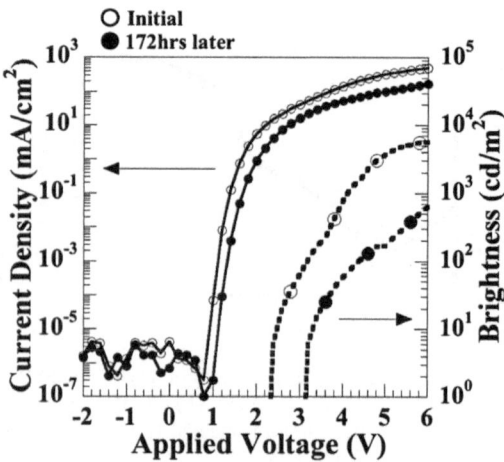

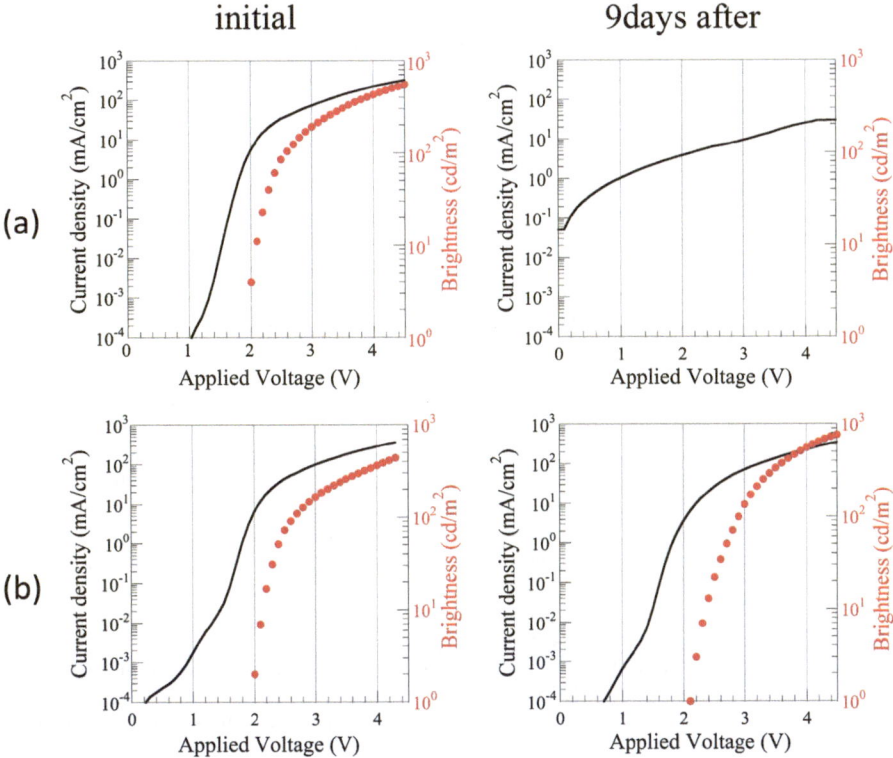

Fig. 2.7 Current density (black line) versus voltage (left ordinate) and luminance (red circles) versus voltage characteristics of initial HOILED devices (left graphs) and (**a**) 9 days in air (upper right graph) and (**b**) in a desiccator with reduced moisture (right bottom graph)

hand, no emission was observed for a conventional OLED after 24 h in air. We also attempted to separate the effects of water and oxygen. Storage lifetimes in air (a) and in a desiccator with reduced moisture (b) were evaluated. Although it was not possible to observe the light in the desiccator with reduced moisture after 9 days of storage, the HOILED in air showed almost the same EL characteristics as in the initial state (Fig. 2.7). This result suggests that the HOILED was only degraded by water molecules.

2.4   Conclusion

By reducing the electron injection barrier at the cathode and introducing the high hole injection interface at the anode, HOILED realized high EL characteristics and high air stability, but it was still considerably below the level required for practical use. Readjustment of the surface energy by a new control method, a new device structure and the development of new materials are essential for the practical use of inverted OLEDs. On this basis, we proposed a novel inverted structure, the iOLED, which incorporated newly developed materials and interface-energy tuning technology to achieve smooth electron injection.

References

1. Aziz, H. et al. Degradation processes at the cathode/organic interface in organic light emitting devices with Mg:Ag cathodes. *Appl. Phys. Lett.* **72**(21), 2642–2644 (1999).
2. Morii, K. et al. Encapsulation-free hybrid organic inorganic light-emitting diodes. *Appl. Phys. Lett.* **89**(18), 183510, (2006).
3. Tokito, S. et al. Metal oxides as a hole-injecting layer for organic electroluminescent device. *J. Phys. D.* **29**, 2750–2753 (1996).
4. Nazeeruddin, M. K. et al. Conversion of light into electricity with trinuclear ruthenium complexes absorbed on textures TiO_2 films. *Helv. Chim. Acta.* **73**(6), 1788–1803 (1990).
5. Altamirano, M. L. et al. Genotoxic studies of vanadium pentaoxide (V2O5) in male mice II, effects in several tissues. *Teratog. Carcinog. Mutagen.* **19**(4), 243–255 (1999).
6. Sessolo, M. et al. Hybrid organic-inorganic light-emitting diodes. *Adv. Mater.* **23**(16), 1829–1845 (2011).
7. Okuya, M. et al. Fabrication of dye-sensitized solar cells by spray pyrolysis deposition (SPD) technique. *J. Am. Chem. Soc.* **164**, 167–172 (2004).
8. Bolink, H. J. et al. Air-stable hybrid organic-inorganic light emitting diodes using ZnO as the cathode. *Appl. Phys. Lett.* **91**, 223501 (2007).
9. Titkov, A. et al. Luminescence up-conversion by Auger process at InP-AlInAs type II interfaces. *Sol. St. Elec.* **37**(4–6), 1041–1044 (1994).
10. Qian, L. et al. Electroluminescence from light emitting polymer/ZnO nanoparticle heterojunctions at sub-bandgap voltages. *Nano Today.* **5**, 384–389 (2010).
11. Saif, A. H. et al. A multilayered polymer light-emitting diode using a nanocrystalline metal-oxide film as a charge-injection electrode. *Adv. Mater.* **19**, 683–687 (2007).
12. Tokmoldin, N. et al. A hybrid inorganic-organic semiconductor light-emitting diode using ZrO_2 as an electron-injection layer. *Adv. Mater.* **21**, 3475–3478 (2009).
13. Morii, K. et al. Enhanced hole injection in a hybrid organic-inorganic light-emitting diode. *Jpn. J. Appl. Phys.* **47**(9 PART1), 7366–7368, (2008).
14. Morii, K. et al. High efficiency and stability in air of the encasulation-free hybrid organic-inorganic light-emitting diode. *Appl. Phys. Lett.* **92**(21), 213304, (2008).
15. Hsieh, T. et al. Surface modification of TiO_2 by a self-assembly monolayer in inverted-type polymer light-emitting devices. *Org. Electro.* **10**, 1626–1631 (2009).
16. Lee, B. R. et al. Highly efficient inverted polymer light-emitting diodes using surface modification of ZnO layer. *Nat. Comm.* **5**, 4840, (2014).
17. Kim, Y. H. et al. Polyethylene imine as an ideal interlayer for highly efficient inverted polymer light-emitting diodes. *Adv. Funct. Mater.* **24**(24), 3808–3814 (2014).

Chapter 3
Interfacial Engineering by Introducing an Interlayer

Abstract As discussed in Chap. 2, it is difficult to inject electrons from the electrode to the organic layer by employing only metal oxides in HOILEDs. To reduce the driving voltage of inverted OLEDs, the EIL in an inverted OLED should be prepared by incorporating metal oxides and an interlayer, as illustrated in Fig. 3.1. Accordingly, the work function (WF) of the EIL surface is reduced, which facilitates the electron injection (the relationship between surface WF and electron injection efficiency will be described in Chap. 4). From this viewpoint, various materials have been studied as interlayers, and improvements of inverted OLED characteristics using such interlayers have been demonstrated.

3.1 Self-Assembled Monolayers

One representative example of an interlayer is self-assembled monolayers (SAMs). Although this type of interlayer was used to promote hole injection in conventional OLEDs [1, 2], it has been reported that the electron injection efficiency can also be improved even in inverted OLEDs. Hsieh et al. employed N-[3-(trimethoxysily) propyl]ethylenediamine (PEDA-TMS) to modify the surface of titanium dioxide (TiO$_2$) with the aim of tuning its conduction band to match the energy level of a high-yellow phenyl-substituted poly(para-phenylenevinylene) copolymer (HY-PPV) for fabricating inverted OLEDs [3]. The EL characteristics of the fabricated inverted OLEDs were significantly improved by modifying the TiO$_2$/ITO cathode using PEDA-TMS (Fig. 3.2). Although no luminescence was observed for the unmodified inverted OLED, the light intensity was 3148 cd/m^2 at 7 V for the PEDA-TMS-modified inverted OLED. The turn-on voltage for the PEDA-TMS-modified inverted OLED was 3.1 V, and the EL efficiency was 0.7 cd/A at 200 mA/cm^2. The enhanced EL efficiency for the PEDA-TMS-modified device was mainly attributed to a decrease in the WF of the cathode, which lowered the injection barrier for electrons to enhance the probability of hole–electron recombination.

© The Author(s) 2020
K. Morii, H. Fukagawa, *Air-Stable Inverted Organic Light-Emitting Diodes*,
SpringerBriefs in Applied Sciences and Technology,
https://doi.org/10.1007/978-3-030-18514-5_3

Fig. 3.1 Multilayered structure of an inverted OLED

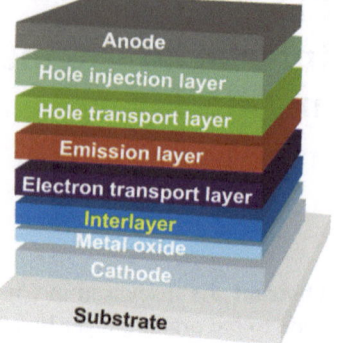

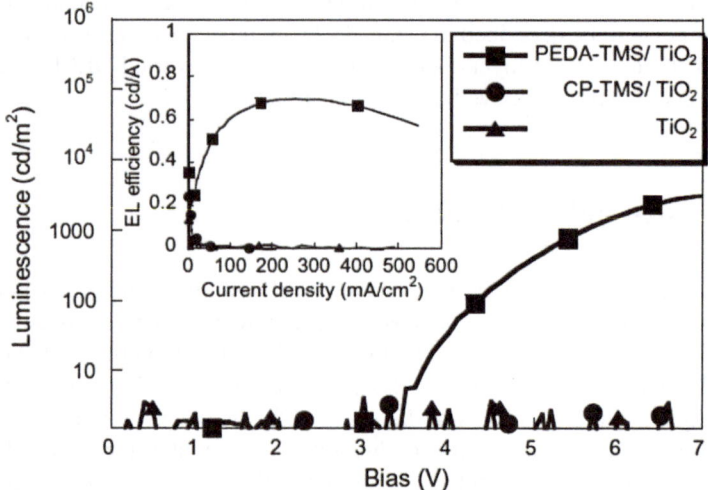

Fig. 3.2 Luminescence–voltage curves of devices with and without self-assembled monolayers. Inset: EL efficiency vs. current density. Reprinted from ref. [3], Copyright 2009, with permission from Elsevier

In addition, J. S. Park et al. reported the effect of SAMs on the ZnO surface in inverted OLEDs [4] (Fig. 3.3a). The SAMs comprised para-substituted benzoic acids functionalised with (i) electron-donating or (ii) electron-withdrawing groups, which led to negative and positive dipoles, respectively (Fig. 3.3b). The charge injection barrier between the conduction band of the ZnO and the lowest unoccupied molecular orbital of the emitting polymer could be easily controlled, resulting in remarkably enhanced electron injection efficiencies, as summarised in Table 3.1. When using the dipole layer with a negatively charged vacuum side, the maximum luminance was limited to 15 cd m^{-2}. With an appropriate dipole orientation, the maximum luminance of the inverted OLED reached 38,200 cd m^{-2} (Table 3.1). It has also been demonstrated that the differences in the device characteristics origi-

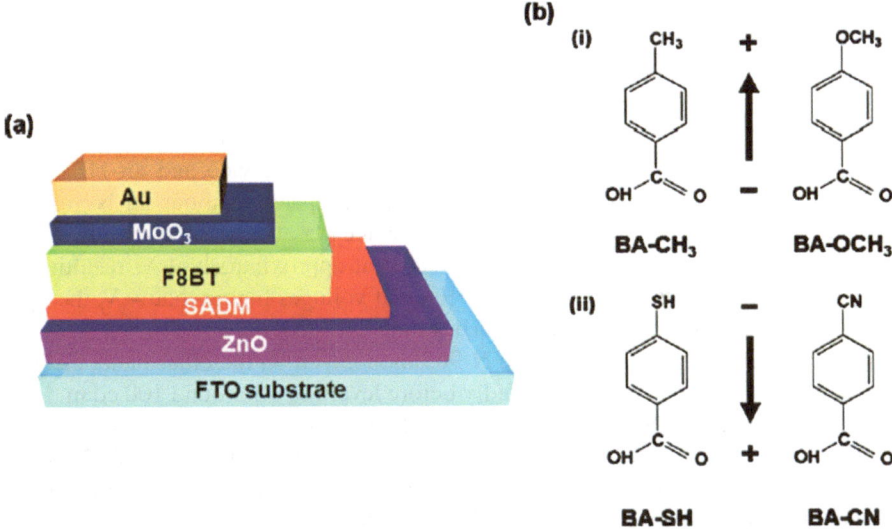

Fig. 3.3 (**a**) Device architecture of an inverted OLED with SAMs (SADMs) on the ZnO surface. (**b**) Chemical structures of carboxylic acid-based SAMs: (**i**) negative dipole molecules: BA-CH3 and BA-OCH3, (**ii**) positive dipole molecules: BA-SH and BA-CN. Reprinted with permission from ref. [4]. Copyright 2010, American Institute of Physics

Table 3.1 Detailed device characteristics of inverted OLEDs without or with SAM modification of ZnO. Reprinted with permission from ref. [4]. Copyright 2010, American Institute of Physics

Device configuration	Maximum luminance (cd/m²) (at voltage)	Maximum luminous efficiency (cd/A) (at voltage)	Turn on voltage (V)
ZnO/F8BT	11,000 (9.4 V)	0.7 (8.0 V)	2.4
ZnO/BA-CH₃	38,000 (9.8 V)	2.8 (9.6 V)	2.2
ZnO/BA-OCH₃	38,200 (10.4 V)	2.1 (10.0 V)	2.4
ZnO/BA-SH	200 (8.8 V)	0.01 (8.6 V)	4.0
ZnO/BA-CN	15 (7.6 V)	0.001 (6.8 V)	4.6

nate from the differences in the surface WF, which strongly depends on the SAM. These data suggest that tuning the electron injection properties of metal oxide interfaces through surface assembly of dipole molecules is an effective approach for fabricating bright and efficient inverted OLEDs.

3.2 Ionic Species

Another approach to improve charge injection into organic semiconductors is the use of ionic species. This strategy has been successfully applied to light-emitting electrochemical cells (LECs). In LECs, the redistribution of ions under an applied bias creates an ionic space charge at the interface, which assists and promotes charge transfer from the organic to the semiconductor or metal and vice versa [5]. H. J. Bolink et al. reported the device characteristics of an inverted OLED, the EIL of which comprised nano-functionalised titanium oxide with a monolayer of an ionic transition metal complex named N-965 [6]. Figure 3.4a shows plots of current density versus voltage for an ITO/TiO$_2$/MEH-PPV/Au device and an ITO/TiO$_2$/ LB-N965/MEH-PPV/Au device. For the device prepared without an N-965 LB monolayer, the current flowing through the device was likely to be mainly a hole current, and electroluminescence was detected at approximately 4 V, resulting in very poor current efficiency (Fig. 3.4b). Even at voltages higher than 4 V, the light output of the device was low, reaching a brightness level of only 5 cd m^{-2} at 10 V. For the inverted OLED prepared with an N-965 monolayer, on the other hand, electro-luminescence started at 3 V, rose rapidly before levelling off around 100 cd m^{-2} and finally reached a maximum of 370 cd m^{-2} at 10 V.

Several poly[(9,9-bis(3′-(N,N-dimethylamino)propyl)-2,7-fluorene)-alt-2,7-(9,9-dioctylfluorene)] (PFN) derivatives have been employed as EIL materials in conventional OLEDs [7–9]. In 2010, poly(9,9′-bis(6″-(N,N,N-trimethylammonium)hexyl) fluorene-alt-co-phenylene) (PFN-BIm4), the chemical structure of which is shown in the inset of Fig. 3.5, was employed as an interlayer material in inverted OLEDs [10]. Figure 3.5 shows the current density–voltage–luminance characteristics of inverted OLEDs with/without PFN-BIm4. The turn-on voltage for the luminance was very low even in the inverted OLED without PFN-BIm4. However, the current onset occurred at a slightly lower voltage, indicating the preferential injection of holes through the MoO$_3$ into the F8BT. A high luminance level exceeding 5000 cd/ m^2 at 10 V, accompanied by an efficacy (Fig. 3.2b) of about 2 cd/A, was obtained. On the other hand, the insertion of a thin PFN-IBm4 film resulted in a drastic change in the current density–voltage–luminance characteristics, as shown in Fig. 3.5a. The current density increased rapidly at about 2.5 V and became almost identical to that of the device without PFN-IBm4 at about 6 V. The onset of the electroluminescence occurred at about 2.6 V, after which it increased rapidly and reached 1000 cd/m^2 at 5 V. The electron injection/transport properties of the polyelectrolyte were reflected by a lower current density and a higher luminance at a lower voltage, resulting in an enhanced efficacy with a maximum value of 3.8 cd/A at 5 V. In 2011, H. Choi et al. reported an inverted OLED using an interlayer material of poly (9,9′-bis(6″-N,N,N-trimethylammoniumhexyl) fluorene-co-alt-phenylene) with bromide counterions (FPQ-Br), the chemical structure of which is similar to that of PFN-IBm4 [11]. As has been discussed, ionic species such as conjugated polyelectrolytes are effective in improving the electron injection efficiency; however, the device characteristics of inverted OLEDs employing ionic species as interlayer materials are sometimes relatively poor compared with those of conventional OLEDs using the same emitting material.

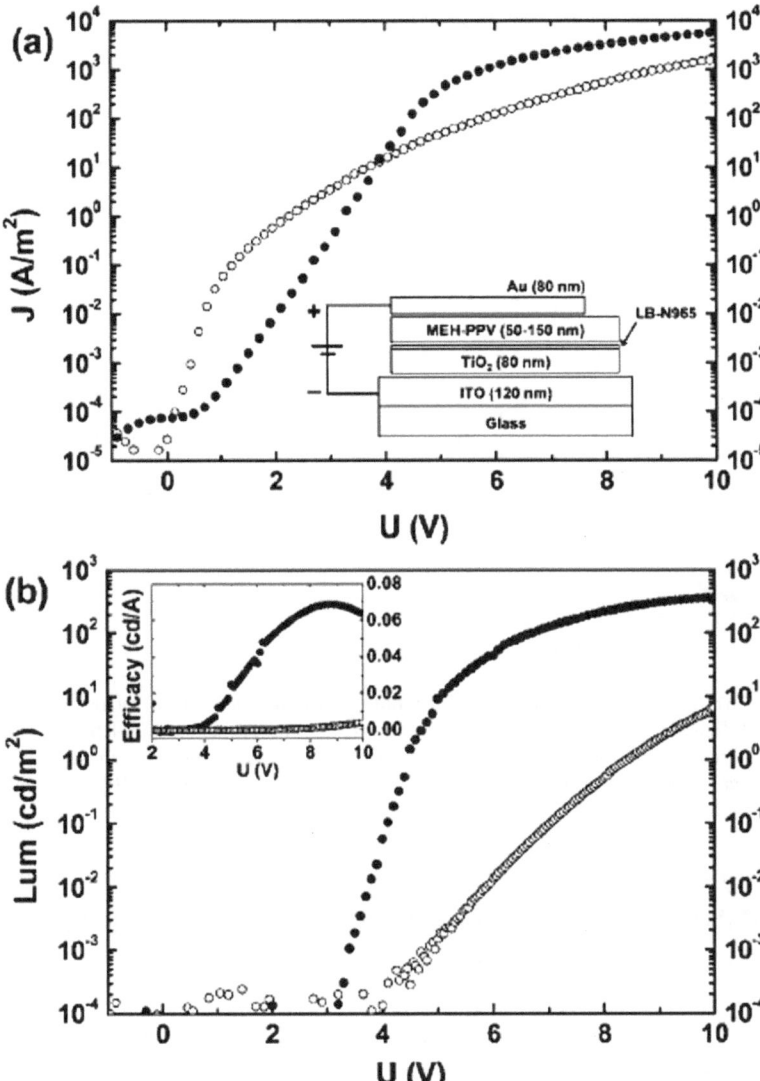

Fig. 3.4 Plots of (**a**) current density and (**b**) luminance versus applied voltage for an ITO/TiO$_2$/MEH-PPV/Au device (open symbols) and an ITO/TiO$_2$/LB-N965/MEH-PPV/Au device (closed symbols). The inset in (**a**) shows a schematic representation of the inverted OLED. The inset in (**b**) shows plots of efficacy versus applied bias for the devices without (open symbols) and with (closed symbols) N-965 monolayer-functionalised TiO$_2$. Reprinted with permission from ref. [6]. Copyright 2009, American Chemical Society

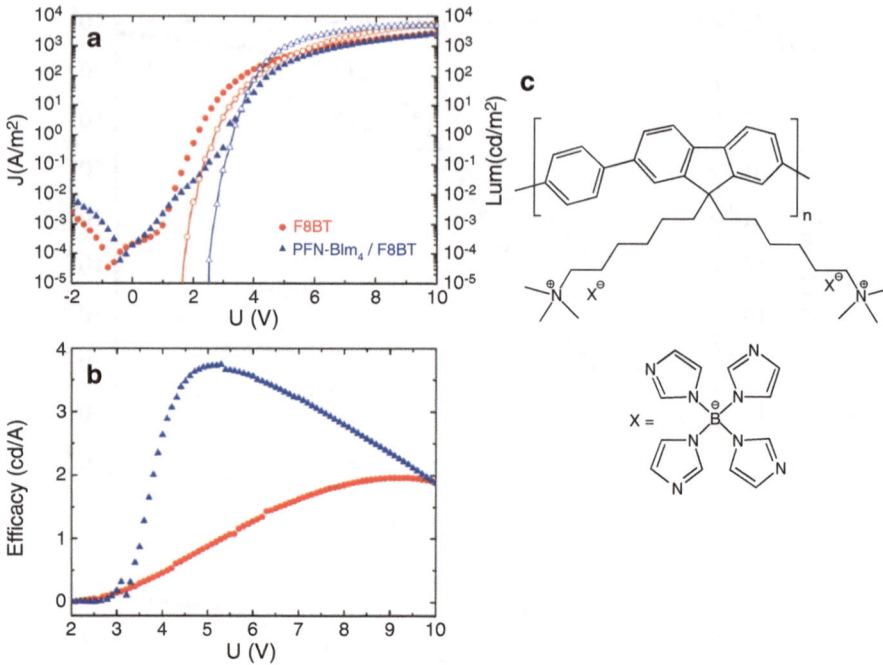

Fig. 3.5 (**a**) Plots of current density (full symbols) and luminance (lines) versus applied voltage for an ITO/ZnO/F8BT/MoO$_3$/Au (red circles) device and an ITO/ZnO/PFN-IBm4/F8BT/MoO$_3$/Au (blue triangles) device. (**b**) Plots of efficacy versus applied bias for these two devices. (**c**) Chemical structure of PFN-IBm4. Reprinted with permission from ref. [10]. Copyright 2010, American Chemical Society

3.3 Polyethylenimine Ethoxylated (PEIE) and Branched Polyethylenimine (PEI)

The report by Zhou et al. in 2012 was a big turning point for the interlayer used in inverted OLEDs [12]. It was reported that an ultrathin layer (1–10 nm) of a polymer containing simple aliphatic amine groups, which are shown in Fig. 3.6, could significantly reduce the surface work function of the EIL. For example, a gradual change in the surface work function by depositing ZnO and PEI on ITO was reported. The work function of ITO was initially about 5.0 eV and was reduced to about 4.0 eV by depositing ZnO. By the physisorption of PEI on ZnO, the work function of the electrode was further reduced to about 3.1 eV, which is comparable to the work function of alkali metals. The electron injection property of the ITO/ZnO/PEI surface was demonstrated to be comparable to that of alkali metals (the relationship between surface work function and carrier injection mechanism will be described in Chap. 4). Although PEI was employed as the EIL in an inverted OLED in 2012, the inverted OLED exhibited a relatively high driving voltage due to the

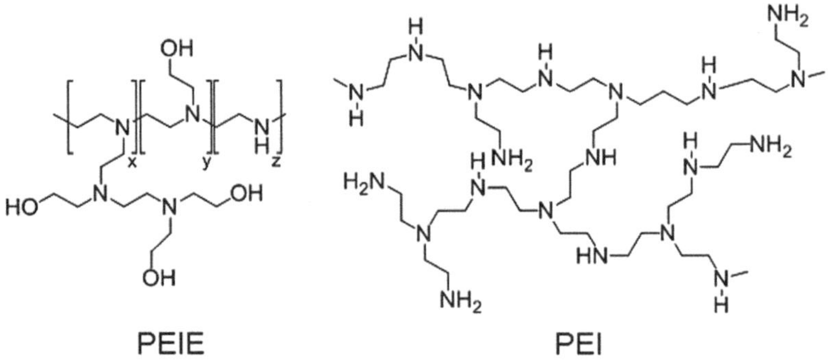

Fig. 3.6 Chemical structures of PEIE and PEI

lack of ZnO between ITO and PEI [13]. Zhou et al. employed PEI as the EIL in a conventional OLED; however, they also did not combine PEI with metal oxides [12]. Inverted OLEDs employing EILs that comprise PEI and metal oxides began to be reported in 2014.

S. Höfle et al. investigated two-step electron injection from a high work function ITO cathode into an efficient and air-stable inverted OLED through PEI on ZnO buffer layers [14]. The inverted OLED was mainly fabricated using solution process except for the hole injection layer and the anode. The configuration of the inverted OLED was ITO/ZnO/PEI/Super Yellow/MoO$_3$/Al. The current density–voltage–luminance (J–V–L) characteristics of the OLEDs are shown in Fig. 3.7a. For devices with an ITO/ZnO cathode, current efficiencies of ~2 cd A^{-1} were achieved (Fig. 3.7b). Due to the reduced injection barrier for electrons, the device current densities increased dramatically when PEI was adopted as the interlayer. The current efficiencies increased to ~22 cd A^{-1} over a much wider luminance range. In essence, the performance of these inverted OLEDs using ITO/ZnO/PEI as cathodes is comparable to that of conventional OLEDs with a configuration of ITO/MoO$_3$/Super Yellow/LiF/Al. Since the inverted OLEDs employing ZnO/PEI as the EIL render the use of reactive materials such as calcium and LiF redundant, the air stability of these inverted OLEDs is expected to be higher than that of conventional OLEDs using reactive materials. In order to investigate the OLED stability in air, they compared the inverted OLEDs with conventional OLEDs comprising ITO/PEDOT:PSS anodes and LiF/Al cathodes. After fabrication and initial characterisation under inert conditions, they investigated the storage lifetime of the non-encapsulated devices under ambient conditions with repetitive testing over several days. At the beginning, the luminance was higher for the conventional OLED ($L = 8600$ cd m^{-2} at $J = 40$ mA cm^{-2}) than for the inverted device ($L = 5900$ cd m^{-2} at $J = 40$ mA cm^{-2}). After only a couple of hours in air, however, the conventional OLED showed strong degradation, i.e., dark spots appeared in the active area. The number and size of these dark spots steadily increased until complete device failure after seven days. Similar device degradation has also been observed for organic solar cells with LiF/Al cathodes [15]. In contrast, over seven days, no dark spots

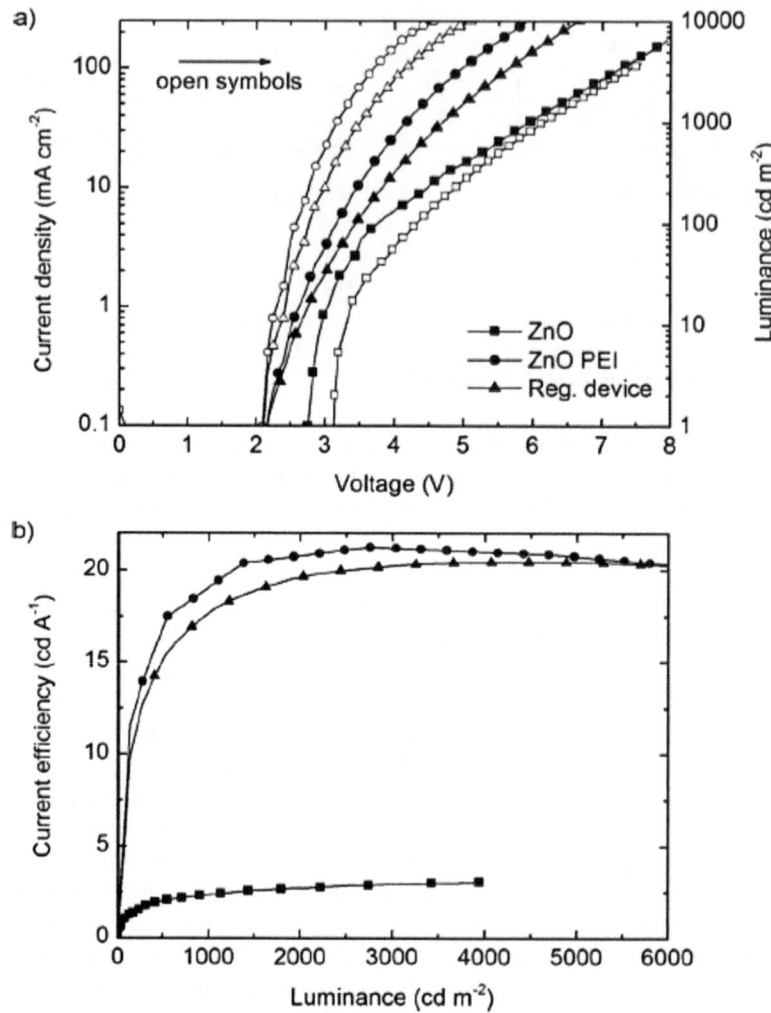

Fig. 3.7 (**a**) J–V–L characteristics of inverted OLEDs comprising ITO/ZnO or ITO/ZnO/PEI cathodes. The ZnO/PEI electron injection layers result in enhanced device currents and consequently higher luminance. (**b**) Plots of device current efficiency versus luminance for OLEDs with ITO/ZnO/PEI cathodes. The devices with ITO/ZnO/PEI cathodes clearly outperform their ITO/ZnO counterparts and match the performance of devices with a regular architecture. Reproduced with permission from ref. [14]. Copyright 2014, Wiley-VCH

were observed at all in the inverted OLEDs, despite constant exposure to ambient conditions. Figure 3.8 shows plots of device voltage and luminance versus time at a constant device current density of 40 mA cm^{-2}. The inverted device exhibited only minor luminance decay from 4900 cd m^{-2} to 4550 cd m^{-2}, and the driving voltage required to obtain the desired current density of 40 mA cm^{-2} remained almost constant over the duration of the investigation.

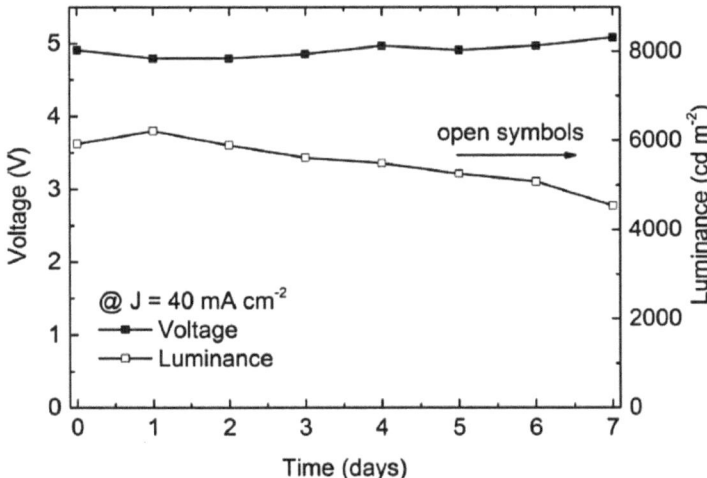

Fig. 3.8 Plots of voltage and luminance versus time for inverted OLEDs stored under ambient conditions at a device current density of $J = 40$ mA cm^{-2}. Reproduced with permission from ref. [14]. Copyright 2014, Wiley-VCH

Y. -H. Kim et al. also reported efficient electron injection from ZnO/PEI and ZnO/PEIE to the emitting polymer [16]. In their report, the electron injection properties of ZnO/PEI and ZnO/PEIE were compared with that of Cs$_2$CO$_3$. The device configuration of the inverted OLEDs was almost the same as that reported by S. Höfle et al. except for the anode material Ag [12]. The dependence of current density on voltage for inverted OLEDs with different interlayers and with no interlayer showed that the devices with PEI and PEIE exhibited lower threshold voltages than the device with Cs$_2$CO$_3$, as shown in Fig. 3.9a. The observed low threshold voltages in devices with PEI and PEIE demonstrate that the interlayer can effectively reduce the electron injection barrier and facilitate electron injection into the emitting layer. Figure 3.9b shows the current efficiency (CE)–voltage characteristics of these inverted OLEDs. The inverted OLED with PEI exhibited a CE of about 13.5 cd A^{-1} and the inverted OLED with PEIE exhibited a CE of about 12 cd A^{-1}. These CEs were much higher than those of the inverted OLEDs with Cs$_2$CO$_3$ ($\approx$ 8 cd A^{-1}) and with no interlayer ($\approx$ 0.08 cd A^{-1}), indicating that PEI and PEIE are excellent at improving electron injection, blocking of holes and blocking of exciton quenching. The effect of PEI on the blocking of exciton quenching was also confirmed from the fact that the photoluminescence intensities of super yellow on ZnO/ PEI became higher with increasing the thickness of PEI as shown in Fig. 3.10. The slightly higher CE of inverted OLEDs with PEI compared with inverted OLEDs with PEIE can be ascribed to the better electron injection capability and reduced exciton quenching of PEI compared with PEIE.

In addition, the authors also reported the effect of the PEI thickness on the device characteristics. It can be seen from Fig. 3.11 that the device characteristics of inverted OLEDs strongly depend on the PEI thickness. As the PEI thickness

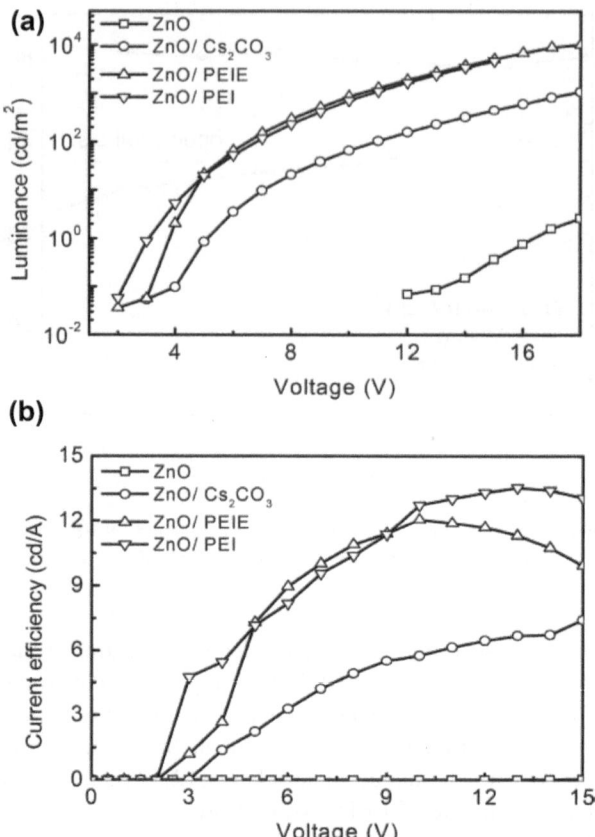

Fig. 3.9 (**a**) Luminance–voltage characteristics of inverted OLEDs with no interlayer, with Cs₂CO₃, PEIE, and PEI. (**b**) Current efficiency–voltage characteristics of inverted OLEDs with no interlayer, with Cs₂CO₃, PEIE, and PEI. Reproduced with permission from ref. [16]. Copyright 2014, Wiley-VCH

increased, the turn-on voltage and operating voltage tended to increase gradually. These trends are due to the increase in the electron injection barrier as the PEI thickness increases. On the other hand, the CE of the device using 8 nm PEI was the highest. The optimum thickness of PEI for inverted OLEDs using Super Yellow was estimated to be about 8 nm. Although PEI can reduce the electron injection barrier and block hole/exciton quenching, employing a thick PEI interlayer can also increase the driving voltage as PEI is an insulating material.

As has been noted, EILs comprising ZnO and PEI have been widely applied in inverted OLEDs using fluorescent polymer emitters since ZnO, PEI and the polymer emitter can be fabricated by solution process. However, the external quantum efficiency (EQE) of OLEDs using fluorescent emitters is relatively low compared with the OLEDs using phosphorescent or thermally activated fluorescent emitters. To demonstrate highly efficient inverted OLEDs, H. Fukagawa et al. fabricated an

Fig. 3.10
Photoluminescence
intensities of super yellow
($\approx$10 nm) on ZnO/PEI
(0 nm, 4 nm, 8 nm, 12 nm,
and 16 nm) Reproduced
with permission from ref.
[16]. Copyright 2014,
Wiley-VCH

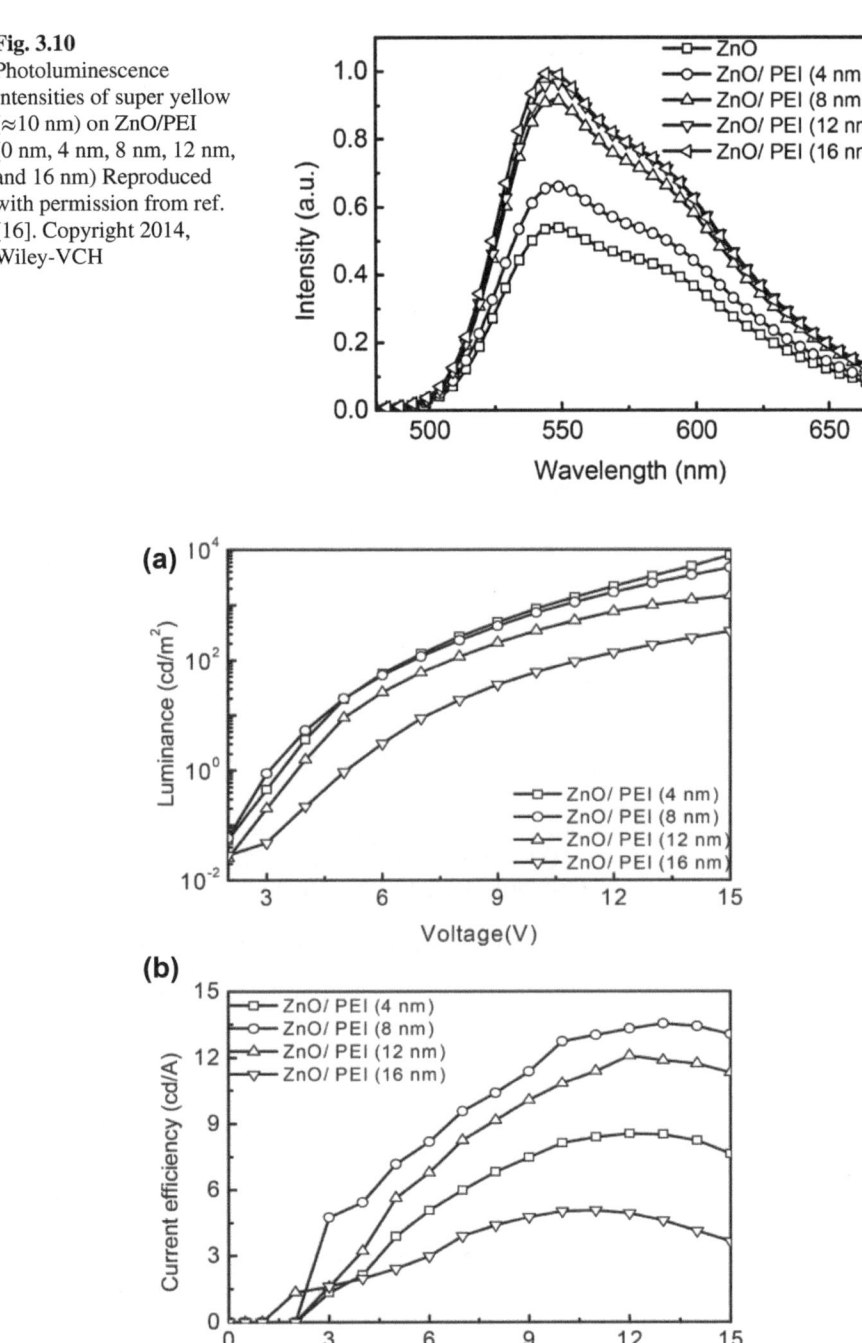

Fig. 3.11 (a) Luminance–voltage characteristics of inverted OLEDs with PEI (4 nm, 8 nm, 12 nm, and 16 nm). (b) Current efficiency–voltage characteristics of inverted OLEDs with PEI (4 nm, 8 nm, 12 nm, and 16 nm). Reproduced with permission from ref. [16]. Copyright 2014, Wiley-VCH

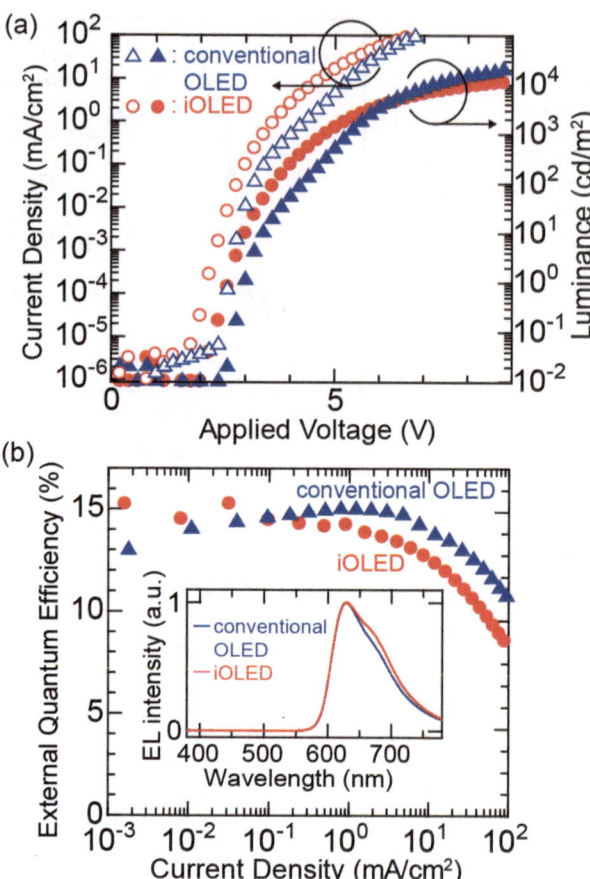

Fig. 3.12 Current density (left, open symbols) and luminance (right, filled symbols)–voltage characteristics of a conventional OLED and two inverted OLEDs. (**b**) External quantum efficiency–current density characteristics of these devices. Inset: electroluminescence spectra of devices at a luminance of 100 cd/m^2

inverted OLED using a red phosphorescent emitter [17]. The device configuration of the inverted OLED is ITO/ZnO/PEI/Bebq$_2$:Ir(piq)$_3$/α-NPD/MoO$_3$/Au, where Ir(piq)$_3$ is tris[1-phenylisoquinolinato-C2,N]iridium(III). In this inverted OLED, ZnO was prepared by sputtering, PEI was prepared by spin coating and the other layers were subsequently deposited by thermal evaporation. In addition to the inverted OLED, a conventional OLED, the configuration of which is ITO/PEDOT:PSS/α-NPD/Bebq$_2$:Ir(piq)$_3$/TPBi/LiF/Al, was fabricated to compare the device characteristics and air stability between the inverted OLED and the conventional OLED. Figure 3.12a shows the J–V–L characteristics of a conventional OLED and two inverted OLEDs; the inverted OLED exhibited similar J–V–L characteristics to the conventional OLED. The driving voltage of the inverted OLED was slightly lower than that of the conventional OLED using TPBi, which is widely used as an electron-transporting layer material. The EQE–J characteristics of the conventional and inverted OLEDs are shown in Fig. 3.12b. The maximum EQEs of the conventional and inverted OLEDs are similar. In both devices, a maximum EQE of over 15% was obtained. It is concluded that by selecting a suitable EIL, almost all carriers injected from the electrodes are detected as $Ir(piq)_3$ emission, even in inverted OLEDs.

In addition to the initial device characteristics, the long-term storage stability of the conventional and inverted OLEDs in air was examined by using a barrier film with a water vapor transmission rate (WVTR) of about 10^{-4} g m^{-2} day^{-1}. Although the short-term air stability of inverted OLEDs was reported by S. Höfle et al. [14], the long-term air stability had not been reported. Both the conventional OLED and the inverted OLED were encapsulated by a glass frame, a barrier film and a UV epoxy resin in a nitrogen atmosphere. The light-emitting areas of the OLEDs, which were kept in air at room temperature and pressure, were observed routinely using an optical microscope. A dc current was applied to the OLEDs during measurements but not under storage conditions. Figure 3.13 shows images of the light-emitting areas of the OLEDs as a function of storage time. The images were obtained without changing the measurement conditions, such as the applied current, diaphragm and exposure time of the optical microscope. Dark-spot formation was clearly observed in the conventional OLED after 15 days of exposure to the atmosphere, and the emitting-area decreased by about half after 103 days. The observed dark-spot formation and emitting-area shrinkage may originate from the degradation/oxidisation of lithium fluoride and/or aluminium, both of which are widely used in conventional OLEDs, and these phenomena are not surprising since it has been proposed that stringent encapsulation (a WVTR of about 10^{-6} g m^{-2} day^{-1}) is necessary for conventional OLEDs [18]. On the other hand, the inverted OLED encapsulated by a barrier film exhibited high storage stability in air; no dark-spot formation or emitting-area shrinkage was observed after 250 days. This is because the inverted OLED was fabricated without using alkali metals or aluminium, resulting in its high storage stability in air.

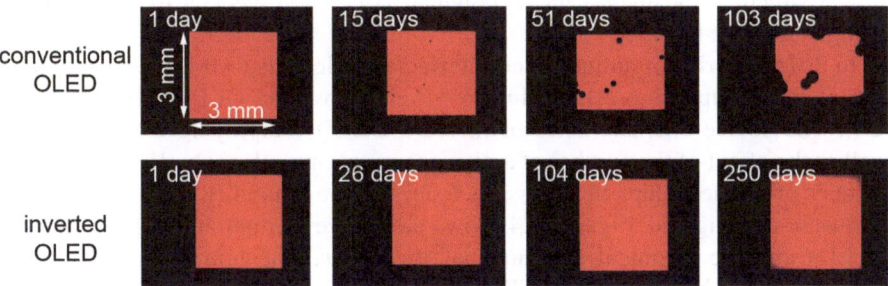

Fig. 3.13 Images of light-emitting areas of OLEDs encapsulated by a barrier film as a function of storage time

As has been noted, PEI and PEIE have been widely used as interlayers to improve the electron injection efficiency of inverted OLEDs. In addition to inverted OLEDs, PEI and PEIE have also been widely applied in quantum-dot LEDs and organic photovoltaics [19, 20, 21]. However, two issues limiting practical use remain. One is that it is essential to control the thickness of PEI to demonstrate ideal inverted OLEDs. The other is that it is essential to prevent accumulation of holes around PEI to demonstrate operationally stable inverted OLEDs [22]. Although seven years have passed since PEI and PEIE were first applied as interlayers, it will take a little more time to apply these interlayers in actual devices, especially for displays that have a bank structure.

Since PEI and PEIE can reduce the surface WF of the electrode significantly, they enabled to inject electron from various electrodes other than ITO. J. H. Park et al. fabricated an inverted OLED on flexible and transparent metallic grid electrodes prepared by evaporative assembly, which is demonstrated by using ZnO/PEI [23]. S. J. Lee et al. fabricated an inverted OLED on highly conductive silver network films on a plastic substrate using ZnO/PEI [24]. Recently, S.-J. Kwon et al. reported demonstrated inverted OLED using graphene electrode [25]. Since it was difficult to inject electron from the graphene electrode to the emitting layer using electron injection layer consists of ZnO/PEIE, they employed n-Type doping with (4-(1,3-dimethyl-2,3-dihydro-1H-benzoimidazol-2-yl)phenyl)   dimethylamine (N-DMBI), resulting in lowered graphene WF.

3.4   Other Amine Derivatives

B. R. Lee et al. reported using amine-based interfacial molecules as the interlayer in inverted OLEDs [26, 27]. In 2014, they applied an amine-based polar solvent treatment to ZnO with a ripple-shaped nanostructure using 2-methoxyethanol (2-ME) and ethanolamine (EA) co-solvents (2-ME + EA) [26]. Although the effect of the surface modification on the electron injection property was not discussed in detail, highly efficient fluorescent inverted OLEDs were demonstrated. The optimised inverted OLED exhibited a luminous efficiency of 61.6 cd A^{-1} and an EQE of 17.8%, which are the highest among polymer-based fluorescent OLEDs that contain a single emitting layer.

In 2015, a series of amine-based interfacial molecules (AIMs), which contain 2–6 amine groups (2–6 N), for highly efficient inverted OLEDs were reported [27]. The device characteristics of inverted OLEDs employing these AIMs, as well as the effect of the number of amine groups of one solvent molecule (the total number of amine groups in AIMs was fixed) on the tunability of the energy barrier, were investigated. Figure 3.14 shows the device configuration of the inverted OLEDs and the chemical structures of the AIMs. The luminance and device efficiencies of the inverted OLEDs with AIMs were significantly higher than those of the reference inverted OLED without AIMs, as shown in Fig. 3.15. This

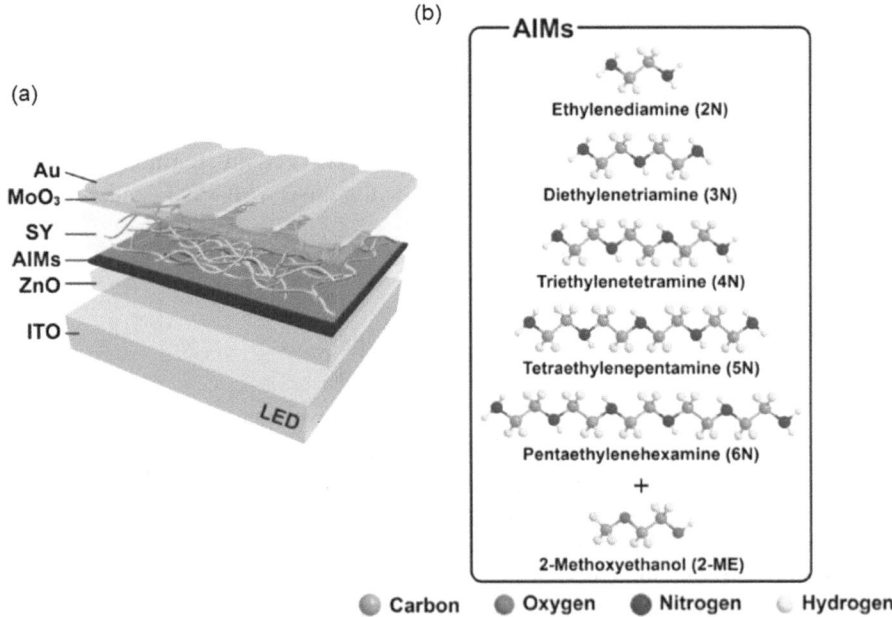

Fig. 3.14 Schematic of an inverted OLED and chemical structures of AIMs. (**a**) Device architecture of the inverted OLED. (**b**) Chemical structures of AIMs, including ethylenediamine (EDA, 2 N), diethylenetriamine (DETA, 3 N), triethylenetetramine (TETA, 4 N), tetraethylenepentamine (TEPA, 5 N), and pentaethylenehexamine (PEHA, 6 N) with 2-methoxyethanol (2-ME). Reproduced with permission from ref. [27]. Copyright 2015, Wiley-VCH

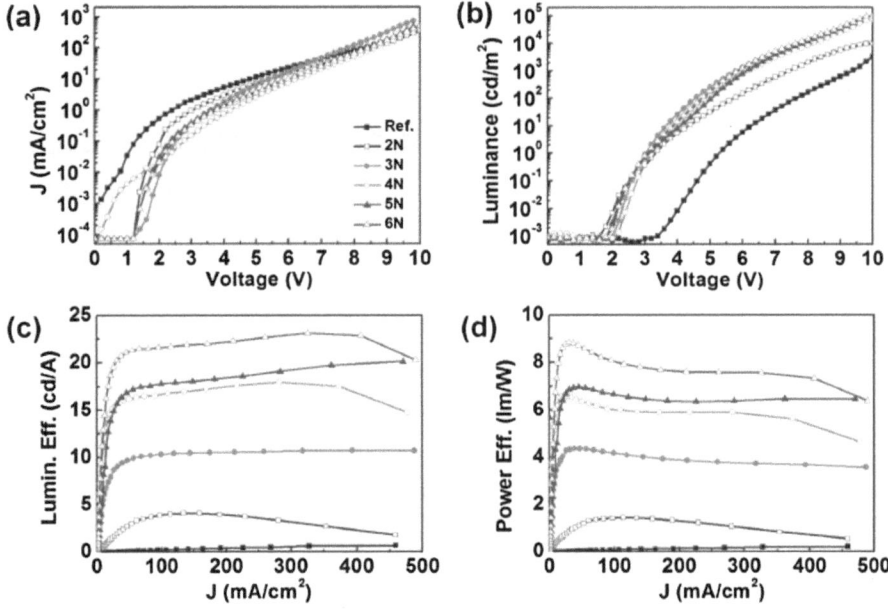

Fig. 3.15 (**a**) J–V, (**b**) L–V, (**c**) LE–J and (**d**) PE–J characteristics of inverted OLEDs. Reproduced with permission from ref. [27]. Copyright 2015, Wiley-VCH

difference in the device characteristics of inverted OLEDs was ascribed to the difference in the electron injection property. In addition, the luminance and device efficiencies of the inverted OLEDs gradually increased with increasing amine group number in AIMs. It was demonstrated that AIMs could significantly reduce the surface work function of ZnO/ITO as in the case with PEIE and PEI (see Chap. 2,3,4).

3.5   Novel Metal Oxides and Electrodes

H. Hosono et al. reported a novel EIL comprising a novel metal oxide a-zinc silicate (a-ZSO) and a-calcium aluminate electride (a-C12A7:e) [28]. A-ZSO has a low work function of 3.5 eV and a high electron mobility of 1 cm^2/(V·s); furthermore, it also forms an ohmic contact with not only conventional cathode materials but also anode materials. A-C12A7:e has an exceptionally low work function of 3.0 eV and was used to improve the electron injection from a-ZSO to an emitting layer. The luminance performance of inverted and normal-type OLEDs with LiF/Al or a-C12A7:e/a-ZSO/Al (ITO) is shown in Fig. 3.16. The turn-on voltages of the inverted OLEDs with a-C12A7:e/a-ZSO were almost the same as or slightly lower than those of the normal-type (conventional) OLED with LiF/Al. The superior electron injection property of the a-ZSOC12A7:e hybrid electron injection layer was demonstrated.

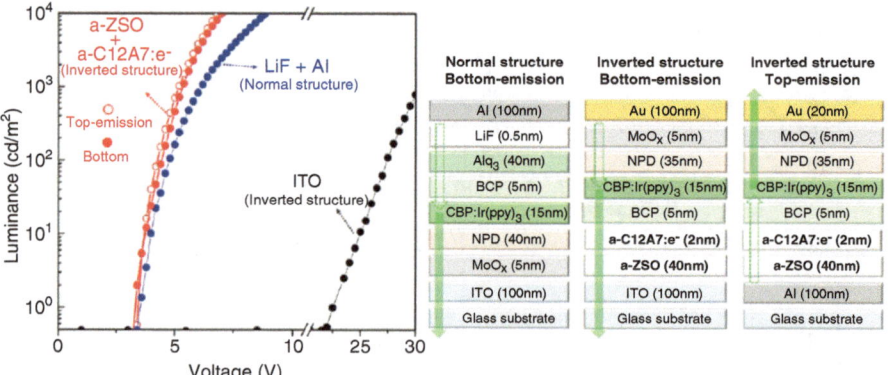

Fig. 3.16 (Left) Luminous intensity–voltage characteristics of OLEDs. (Right) Device stacking structure. Reproduced with permission from ref. [28]

3.6   Solution-Processed Boron Compounds for Operationally Stable Inverted OLEDs

Although many efficient inverted OLEDs were reported, inverted OLEDs have rarely been used in actual devices such as displays for the following two reasons. One reason is that the development of interlayers that can prevent the decrease in brightness caused by inverted OLED operation is lacking. The other reason is that the applicability of inverted OLEDs in actual devices has not been verified, as typified by amine derivatives. However, H. Fukagawa et al. reported a long-lived flexible display employing efficient and stable inverted OLEDs [29]. They found that a thin film of boron compounds fabricated via solution process, which was formed as an interlayer between ZnO and the electron transport layer, could significantly reduce the driving voltage of inverted OLEDs. An energy level diagram of the inverted OLEDs is illustrated in Fig. 3.17a, and the interlayer-dependent device characteristics of the inverted OLEDs are shown in Fig. 3.17c–e; Fig. 3.17b shows the interlayer materials used in the inverted OLEDs. The driving voltage of the inverted OLED comprising a spin-coated tris-[3-(3-pyridyl)mesityl]borane (3TPYMB) interlayer was much lower than that comprising an evaporated

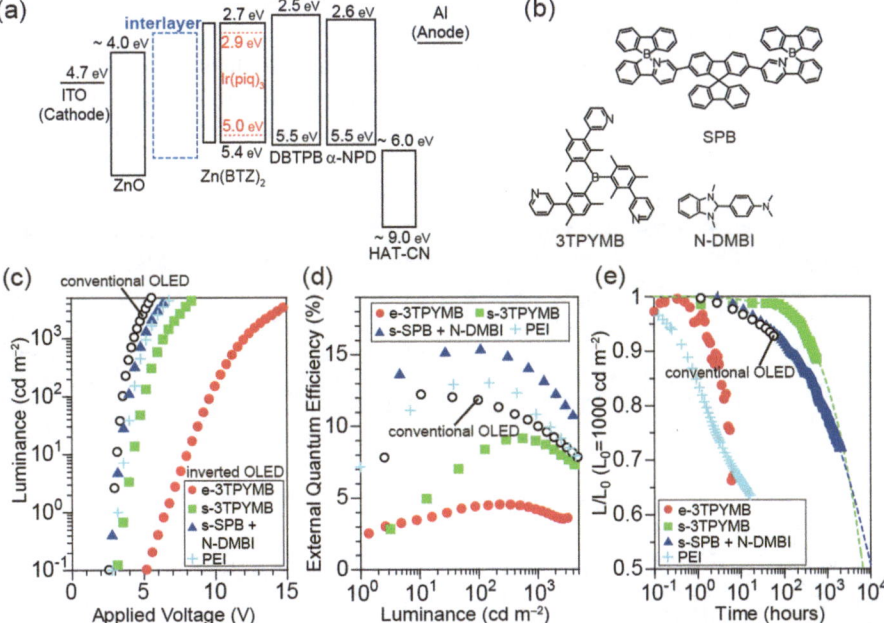

Fig. 3.17 (**a**) Energy level diagram of inverted OLEDs. (**b**) Molecular structures of the materials used as the interlayer in inverted OLEDs. (**c**) Luminance–voltage and (**d**) EQE–luminance characteristics of a conventional OLED (open symbols) and inverted OLEDs (closed symbols and plus symbols) using either an evaporated interlayer (e-3TPYMB) or solution-processed interlayers (s-3TPYMB, s-SPB + N-DMBI, PEI). (**e**) Luminance–time characteristics of devices under a constant dc current with an initial luminance of 1000 cd m⁻²

3TPYMB interlayer, as shown in Fig. 3.17c, suggesting improved electron injection. As a result of the improved electron injection, the inverted OLED employing solution-processed 3TPYMB exhibited a higher EQE and a longer operational lifetime, as shown in Fig. 3.17d and e. Thus, solution-processed boron-containing compounds are considered to be a suitable interlayer material for demonstrating operationally stable inverted OLEDs.

Although operationally stable inverted OLEDs can be achieved by employing boron compounds, the driving voltages of these inverted OLEDs were much higher than that of a conventional OLED using the same emitting layer, as shown in Fig. 3.17c. To lower the driving voltage, 4-(1,3-dimethyl-2,3-dihydro-1*H*-benzoimidazol-2-yl)phenyl)dimethylamine (N-DMBI) [30], which is an air-stable n-type dopant, was added to the interlayer. According to Fig. 3.17c, d and e, the device characteristics of inverted OLEDs employing SPB and N-DMBI were comparable to those of a conventional OLED. In particular, the operational stability of the optimised inverted OLED was much higher than that of an inverted OLED employing PEI.

The reason why solution processing is essential for achieving efficient and stable inverted OLEDs was also examined. ZnO was partially dissolved by the solvent used for spin coating of the interlayer, resulting in deep penetration of ZnO into the interlayer. The electron injection efficiency was improved by the interpenetrated ZnO because electrons could be effectively transported near the interface between the interlayer and the electron-transport layer.

The applicability of the solution-processed interlayer to displays and the air stability of the inverted OLED-based devices were examined by fabricating two flexible displays as shown in Fig. 3.18. One flexible display employed inverted

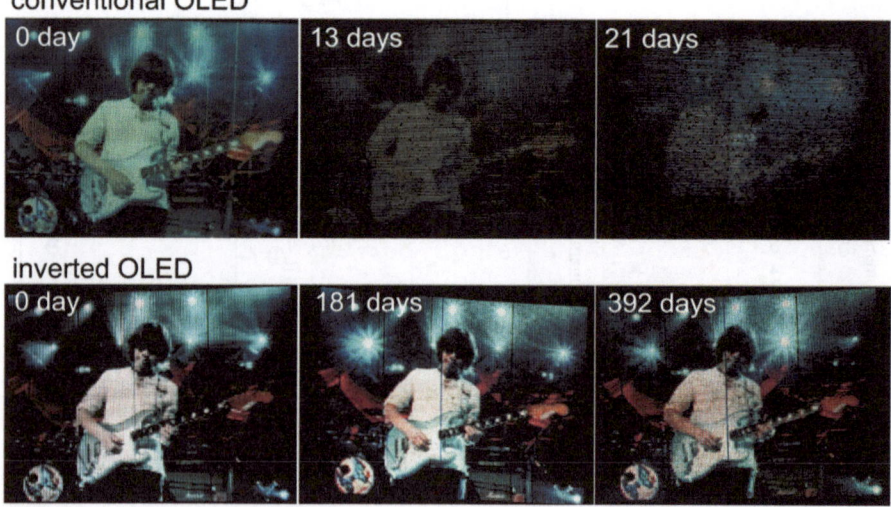

Fig. 3.18 Photographs of inverted OLED- and conventional OLED-based flexible displays in operation as a function of storage time (image source: ITE/ARIB HDTV Test Materials - Second Edition)

OLEDs using ZnO/solution-processed SPB:N-DMBI as the EIL, and the other one employed conventional OLEDs using LiF as the EIL. The air stability of the two flexible displays was examined by storing them in air. On the conventional OLED-based flexible display, dark-spot formation and emitting-area shrinkage rapidly progressed, and the moving image could not be observed after 21 days. On the contrary, a clear moving image was observed on the inverted OLED-based flexible display even after 392 days. Therefore, the issues with inverted OLEDs such as the operational lifetime and applicability in actual devices can be solved by employing solution-processed boron compounds as the interlayer.

3.7 Conclusion

In this chapter, interlayers that can improve the electron injection efficiency are introduced. Although metal oxides can reduce the electron injection barrier between the organic layer and ITO, which is widely used as the cathode in inverted OLEDs, it is difficult to inject electrons from ITO to the organic layer directly. In the early stages of interlayer development, SAMs and ionic species were intensively used. After the reports about PEIE and PEI, amine derivatives were widely used as the interlayer in inverted OLEDs. More recently, it has been demonstrated that electrides can be used effectively as the interlayer. Efficient and stable inverted OLEDs have been demonstrated by employing solution-processed boron compounds as the interlayer, and higher air stability of inverted OLEDs compared with conventional OLEDs has also been demonstrated in many reports.

References

1. Choi, B., Rhee, J. & Lee, H. H. Tailoring of self-assembled monolayer for polymer light-emitting diodes. *Appl. Phys. Lett.* **79**, 2109–2111 (2001).
2. Nicht, S., Kleemann, H., Fischer, A., Leo, K. & Lüssem, B. Functionalized p-dopants as self-assembled monolayers for enhanced charge carrier injection in organic electronic devices. *Org. Electro.* **15**, 654–660 (2014).
3. Hsieh, S.-N. et al. Surface modification of TiO2 by a self-assembly monolayer in inverted-type polymer light-emitting devices. *Org. Electro.* **10**, 1626–1631 (2009).
4. Park, J. S. et al. Efficient hybrid organic-inorganic light emitting diodes with self-assembled dipole molecule deposited metal oxides. *Appl. Phys. Lett.* **96**, 243306 (2010).
5. Yu, G., Gao, J., Hummelen, J. C., Wudl, F. & Heeger, A. J. Polymer Photovoltaic Cells - Enhanced Efficiencies Via a Network of Internal Donor-Acceptor Heterojunctions. *Science* **270**, 1789–1791 (1995).
6. Bolink, H. J. et al. Molecular ionic junction for enhanced electronic charge transfer. *Langmuir* **25**, 79–83 (2009).
7. Huang, F. et al. High-efficiency, environment-friendly electroluminescent polymers with stable high work function metal as a cathode: green- and yellow-emitting conjugated polyfluorene polyelectrolytes and their neutral precursors. *J. Am. Chem. Soc.* **126**, 9845–9853 (2004).

8. Wu, H. et al. Efficient Electron Injection from a Bilayer Cathode Consisting of Aluminum and Alcohol-/Water-Soluble Conjugated Polymers. *Adv. Mater.*. **16**, 1826–1830 (2004).
9. Hoven, C. V. et al. Electron injection into organic semiconductor devices from high work function cathodes. *Proc Natl Acad Sci U S A* **105**, 12730–12735 (2008).
10. Bolink, H. J., Brine, H., Coronado, E. & Sessolo, M. Ionically Assisted Charge Injection in Hybrid Organic–Inorganic Light-Emitting Diodes. *ACS Appl. Mater. Interfaces* **2**, 2694–2698 (2010).
11. Choi, H. et al. Combination of titanium oxide and a conjugated polyelectrolyte for high-performance inverted-type organic optoelectronic devices. *Adv. Mater.* **23**, 2759–2763 (2011).
12. Zhou, Y. et al. A universal method to produce low-work function electrodes for Org. Electro.. *Science* **336**, 327–332 (2012).
13. Chen, J. et al. Solution-processable small molecules as efficient universal bipolar host for blue, green and red phosphorescent inverted OLEDs. *J. Mater. Chem.* **22**, 5164 (2012).
14. Höfle, S., Schienle, A., Bruns, M., Lemmer, U. & Colsmann, A. Enhanced electron injection into inverted polymer light-emitting diodes by combined solution-processed zinc oxide/polyethylenimine interlayers. *Adv. Mater.* **26**, 2750–2754, (2014).
15. Hau, S. K. et al. Air-stable inverted flexible polymer solar cells using zinc oxide nanoparticles as an electron selective layer. *Appl. Phys. Lett.* **92**, 253301 (2008).
16. Kim, Y.-H. et al. Polyethylene imine as an ideal interlayer for highly efficient inverted polymer light-emitting diodes. *Adv. Funct. Mater.* **24**, 3803–3814 (2014).
17. Fukagawa, H. et al. Highly efficient and air-stable inverted organic light-emitting diode composed of inert materials. *Appl. Phys. Express* **7**, 082104 (2014).
18. Charton, C. et al. Development of high barrier films on flexible polymer substrates. *Thin Solid Films* **502**, 99–103 (2006).
19. Chen, H. C., Lin, S. W., Jiang, J. M., Su, Y. W. & Wei, K. H. Solution-processed zinc oxide/polyethylenimine nanocomposites as tunable electron transport layers for highly efficient bulk heterojunction polymer solar cells. *ACS Appl. Mater. Interfaces* **7**, 6273–6281 (2015).
20. Kim, H. H. et al. Inverted quantum dot light emitting diodes using polyethylenimine ethoxylated modified ZnO. *Sci. Rep.* **5**, 8968 (2015).
21. Kim, D. et al. Polyethylenimine Ethoxylated-Mediated All-Solution-Processed High-Performance Flexible Inverted Quantum Dot-Light-Emitting Device. *ACS Nano* **11**, 1982–1990 (2017).
22. Stolz, S., Zhang, Y., Lemmer, U., Hernandez-Sosa, G. & Aziz, H. Degradation Mechanisms in Organic Light-Emitting Diodes with Polyethylenimine as a Solution-Processed Electron Injection Layer. *ACS Appl. Mater. Interfaces* **9**, 2776–2785 (2017).
23. Park, J. H. et al. Flexible and transparent metallic grid electrodes prepared by evaporative assembly. *ACS Appl. Mater. Interfaces* **6**, 12380–12387 (2014).
24. Lee, S. J. et al. A roll-to-roll welding process for planarized silver nanowire electrodes. *Nanoscale* **6**, 11828–11834 (2014).
25. Kwon, S. J. et al. Solution-Processed n-Type Graphene Doping for Cathode in Inverted Polymer Light-Emitting Diodes. *ACS Appl. Mater. Interfaces* **10**, 4874–4881 (2018).
26. Lee, B. R. et al. Highly efficient inverted polymer light-emitting diodes using surface modifications of ZnO layer. *Nat. Commun.* **5**, 4840 (2014).
27. Lee, B. R. et al. Amine-Based Interfacial Molecules for Inverted Polymer-Based Optoelectronic Devices. *Adv. Mater.* **27**, 3553–3559 (2015).
28. Hosono, H., Kim, J., Toda, Y., Kamiya, T. & Watanabe, S. Transparent amorphous oxide semiconductors for Org. Electro.: Application to inverted OLEDs. *Proc Natl Acad Sci U S A* **114**, 233–238 (2017).
29. Fukagawa. H. et al. Long-Lived flexible displays employing efficient and stable inverted organic light-emitting diodes. *Adv. Mater.* **30**, 1706768 (2018).
30. Wei, P., Oh, J. H., Dong, G., & Bao, Z. Use of a 1H-Benzoimidazole derivative as an n-type dopant and to enable air-stable solution-processed n-channel organic thin-film transistors. *J. Am. Chem. Soc.* **132**, 8852–8853 (2010).

Chapter 4
Carrier Injection Mechanism

Abstract Both hole and electron injection layers are commonly used in recent OLEDs to reduce the injection barrier between electrodes and organic layers. This injection barrier originates from the energy difference between the work function (WF) of the electrode and the energy level of the organic layer. For instance, the hole injection barrier is defined as the energy difference between the Fermi level of the anode and the highest occupied molecular orbital (HOMO) level of the organic layer on the anode, as shown in Fig. 4.1a. Thus, an ideal hole injection material is the material that can make the surface WF of the anode larger (Fig. 4.1b). On the other hand, an ideal electron injection material is the material that can make the surface WF of the cathode smaller (Fig. 4.1c).

Since the relationship between the WF of the electrode and the energy level of the organic layer is important in determining the carrier injection barrier, both material-dependent OLED characteristics and energy diagrams between the electrode and the injection materials have been extensively studied using a Kelvin probe (KP) and by ultraviolet photoelectron spectroscopy (UPS). In particular, hole injection materials have been intensively studied since the anode and the cathode of a conventional OLED are the bottom electrode and the top electrode, respectively. Alkali metals such as Li, Cs and Ba have been used as EILs in conventional OLEDs [1, 2]; however, it was difficult to investigate the electronic structures at the organic layer/EIL/cathode interfaces since alkali metals can react with both the organic layer and the cathode [3]. On the other hand, the electronic structures at the anode/ HIL interface in conventional OLEDs, which can be easily investigated by KP and UPS, have been intensively studied [4]. Similarly, the electronic structures at the cathode/EIL interface in inverted OLEDs have recently been studied. In addition, the electronic structures at the organic layer/HIL/anode interfaces in inverted OLEDs, which are as complicated as the electronic structures at organic layer/EIL/ cathode interfaces, have also become the subject of investigation.

In this chapter, we first explain the basic carrier injection mechanism and related electronic structures using examples of anode/HIL interfaces in conventional OLEDs that have been widely studied. Then, we introduce the electron/hole injection mechanism related to inverted OLEDs.

© The Author(s) 2020
K. Morii, H. Fukagawa, *Air-Stable Inverted Organic Light-Emitting Diodes*,
SpringerBriefs in Applied Sciences and Technology,
https://doi.org/10.1007/978-3-030-18514-5_4

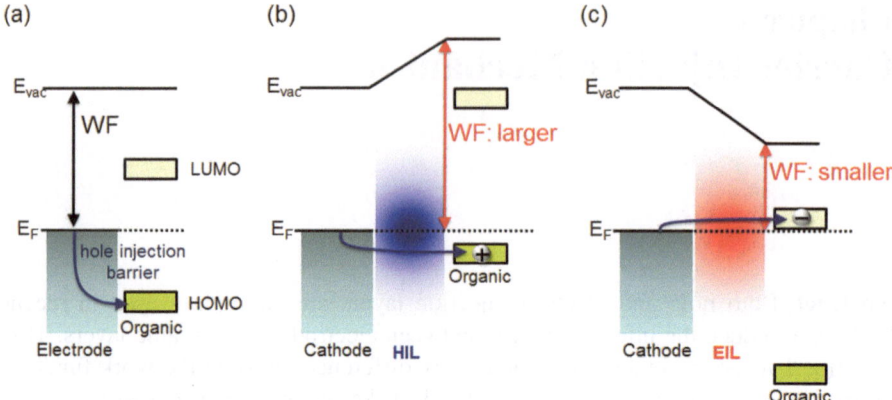

Fig. 4.1 Schematic diagrams of energy level alignment for (**a**) a typical electrode–organic interface, (**b**) an ideal electrode–organic interface for hole injection and (**c**) an ideal electrode–organic interface for electron injection

4.1 Basic Carrier Injection Mechanism and Related Electronic Structures

As shown in Fig. 4.1b, a high WF surface is essential for hole injection. Thus, materials with a high WF have been used to ensure effective hole injection into the organic semiconductor in conventional OLEDs. Transparent conducting oxides, typified by indium tin oxide (ITO), are widely used in bottom-emission or transparent OLEDs. Although the WF of ITO can be up to 5 eV [5], a hole injection barrier of about 0.5 eV still exists between the ITO anode and most hole-transporting materials (HTMs) (Fig. 4.2) [6]. In the case of energy level alignment, where the vacuum levels are aligned, the hole injection barrier can be over 0.5 eV, since the ionisation energy (IE) of an HTM, typified by N,N′-diphenyl-N,N′-bis(1-naphthyl phenyl)-1,1′-biphenyl-4,4′-diamine (NPB), is about 5.5 eV. To minimise the hole injection barrier from the anode to the HTMs, hole injection materials (HIMs), which can enlarge the surface WF, are often employed. The molecular structures of common HIMs are shown in Fig. 4.3.

A typical example of an HIM is tetrafluoro-tetracyano-quinodimethane (F4-TCNQ), which is a strong acceptor. A reduction in the driving voltage of OLEDs by an F4-TCNQ-doped HTL and a reduction in the hole injection barrier by F4-TCNQ doping have been reported in the literature [7–9]. Pfeiffer et al. reported that the conductivity and hole injection property of phthalocyanine films were improved by adding F4-TCNQ [7]. In addition, doping F4-TCNQ into an HIM comprising TDATA was shown to not only reduce the driving voltage of OLEDs but also improve the OLED efficiency [8]. The electronic structure of an F4-TCNQ-doped zinc phthalocyanine (ZnPc) film, which was studied by UPS and inverse photoemission spectroscopy, was also reported by Gao et al. [9]. In their study, the hole injection barrier between an Au electrode and ZnPc was studied by UPS while

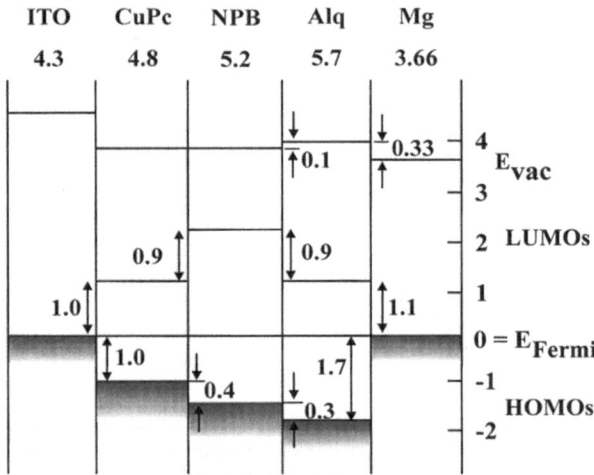

Fig. 4.2 Schematic diagram of energy level alignment detected by UPS for an OLED with the configuration of ITO/CuPc/NPB/Alq/Mg. Reprinted with permission from ref. [6]. Copyright 1999, American Institute of Physics

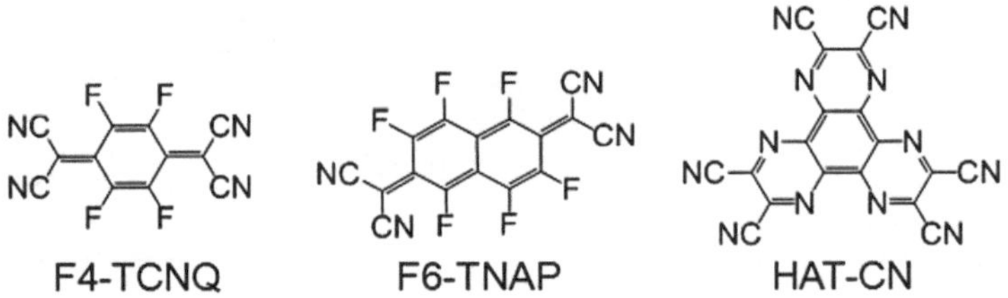

Fig. 4.3 Examples of the molecular structures of HIMs, which are characterised as strong acceptor materials with deep LUMO levels

increasing the thickness of the ZnPc film on the clean Au surface, as shown in Fig. 4.4. The energy difference between the Fermi level of Au and the HOMO level of ZnPc was 0.9 eV, corresponding to a relatively large hole injection barrier. This large hole injection barrier was reduced to 0.18 eV when using an F4-TCNQ-doped ZnPc film. The observed significant reduction in the hole injection barrier was ascribed to efficient charge transfer due to the excellent match between the IE of ZnPc and the electron affinity (EA) of F4-TCNQ, which was detected by inverse photoemission spectroscopy [10]. This charge transfer is defined as p-type doping in an organic semiconductor, and thus the conductivity of the organic semiconductor film can also be improved. Gao et al. also reported that p-type doping was effective for improving the hole injection property and conductivity of an HTM film using NPB, which is a typical HTM (See Fig. 4.5) [10]. Such p-doped HIMs were

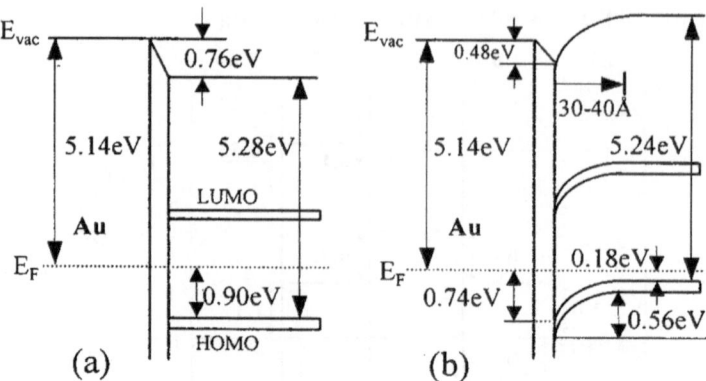

Fig. 4.4 Electronic structures of (a) an undoped ZnPc/Au interface and (b) a ZnPc:3%F4-TCNQ/Au interface determined by UPS and inverse photoemission measurements. Reprinted with permission from ref. [9]. Copyright 2001, American Institute of Physics

Fig. 4.5 Schematic of the electronic structures of NPB and F4-TCNQ. The IE and electron affinity of both materials are shown and were determined by UPS and inverse photoemission measurements. The arrow indicates the presumed electron transfer from the NPB HOMO to the F4-TCNQ LUMO. Reprinted with permission from ref. [10]. Copyright 2003, American Institute of Physics

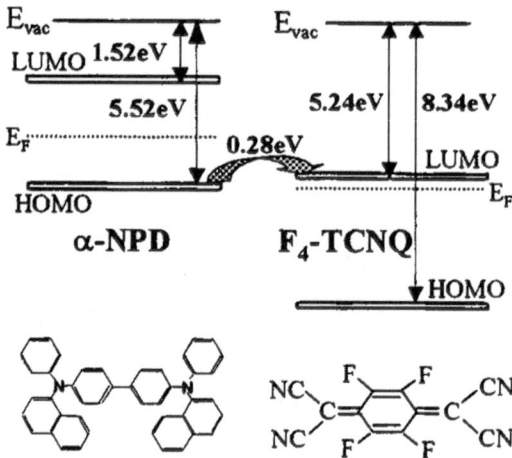

shown to be effective in not only reducing the driving voltage of OLEDs but also improving the efficiency and operational stability of OLEDs [11, 12].

The hole injection barrier at the metal/organic layer interface can also be reduced by preparing a thin film of F4-TCNQ on the electrode surface. Koch et al. carefully examined the formation of the interface dipole at an Au/F4-TCNQ monolayer interface [13]. During the evaporation of an organic layer on a metal surface, the substrate WF of typical metals generally decreases since the part of the electron cloud tailing into vacuum is pushed back by repulsion from the electron cloud in the adsorbate [14]. In the case of F4-TCNQ on a metal substrate, on the other hand, F4-TCNQ formed a charge transfer complex on an Au substrate, resulting in an

Fig. 4.6 Schematic energy level diagram of α-sexithienyl (6P) on Au with pre-adsorbed F4-TCNQ, mainly investigated by UPS. Reprinted with permission from ref. [13]. Copyright 2005, American Physical Society

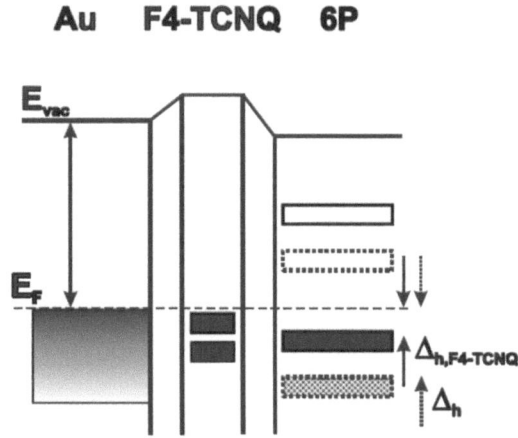

increased substrate WF [13] (Fig. 4.6). An electric dipole layer with a negatively charged vacuum side was formed by the electron transfer from Au to the adsorbed F4-TCNQ, which was detected by UPS and X-ray photoelectron spectroscopy. The hole injection barrier between the F4-TCQN-modified Au substrate and the p-sexiphenyl deposited on F4-TCNQ was much smaller than that between the Au substrate and the α-sexithienyl (6P). Similar electronic structures of F4-TCNQ on several substrates such as metal electrodes [15] and diamond [16], in which electrons transfer from the substrate to the adsorbed F4-TCNQ, have been reported. As discussed earlier, F4-TCNQ is one of the most well-known acceptor materials, and phenomena caused by its strong acceptor properties have been intensively investigated. More recently, a similar molecule, 1,3,4,5,7,8-hexafluorotetracyanonaphthoq uinodimethane (F6-TNAP), the acceptor properties of which are stronger than those of F4-TCNQ, has also been reported [17].

Besides F4-TCNQ, another commonly used acceptor material in OLEDs is 2,3,6,7,10,11-hexacyano-1,4,5,8,9,12-hexaazatriphenylene (HAT-CN), which was reported by LG Chem, Ltd. in 2004. HAT-CN is a strong electron-withdrawing molecule because of its six nitrile groups. The electronic structures related to HAT-CN have been investigated along with its application to OLEDs. Kim et al. reported that the hole injection barrier at an Au/NPB interface was significantly reduced by inserting HAT-CN between the Au substrate and NPB [18] (Fig. 4.7). Although the surface WF of a thick HAT-CN film deposited on Au was about 6.1 eV, it remained a conductive molecular layer. It was found that the lowest unoccupied molecular orbital (LUMO) level of HAT-CN was very close to the Fermi level of Au, and the HOMO level of NPB was located only about 0.3 eV below the Fermi level of Au at this molecular interface. It was suggested that this arrangement would allow the electrons from the NPB HOMO to be easily excited to the HAT-CN LUMO, forming a charge carrier generation interface. As HAT-CN is also an electrically/chemically stable material, it is commonly used for demonstrating operationally stable OLEDs [19].

Fig. 4.7 Energy level diagram at an NPB/ HAT-CN/Au interface. Reprinted with permission from ref. 18. Copyright 2009, American Institute of Physics

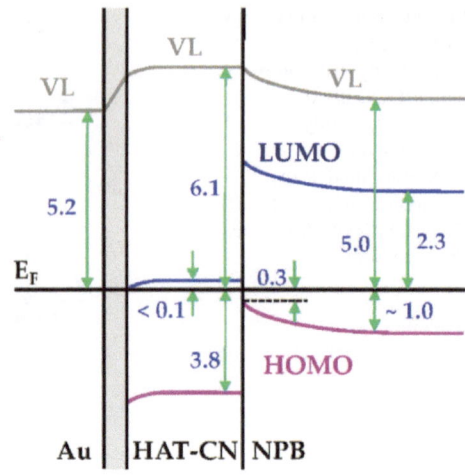

In addition to organic compounds, some metal oxides have also been used to enlarge the metal WF. The first reported metal oxide HILs were vanadium oxide (VOx), molybdenum oxide (MoOx) and ruthenium oxide (RuOx), which were reported by Tokito et al. in 1996 [20]. The driving voltages of OLEDs can be significantly reduced by using metal oxides as HILs. In particular, MoOx is widely used owing to its good hole injection property and relatively low sublimation temperature. The carrier injection mechanisms and related electronic structures of MoOx have been investigated. Matsushima et al. reported that a device with a 0.75 nm-thick MoO_3 HIM could achieve ohmic hole injection from the ITO anode to NPB [21]. The ohmic hole injection was attributed to the electron transfer from ITO and NPB to MoO_3. Kröger et al. reported a detailed electronic structure at ITO/MoO_3/NPB interfaces using UPS, as shown in Fig. 4.8 [22]. The observed phenomena are similar to those in the case of Au/HAT-CN/NPB interfaces. The enhancement in hole injection by inserting MoO_3 was concluded to be due to a very high WF and the deep-lying unoccupied states, i.e., the large EA, of the transition metal oxide. Kröger et al. also provided solid experimental evidence for hole injection via electron extraction from the NPB HOMO through the MoO_3 conduction band. More recently, an enhancement in hole injection from graphene to an HTL has also been demonstrated by inserting MoO_3 between a graphene anode and an HTL [23]. The surface WF of graphene/MoO_3 reached 6.6 eV, resulting from the electron transfer from graphene to MoO_3 (Fig. 4.9).

So far we have outlined the hole injection mechanisms and the characteristics of related materials. Electrodes with high WF and suitable acceptor materials are essential for effective hole injection. Acceptor materials such as F4-TCNQ, HAT-CN and metal oxides are used not only for enlarging surface WF but also for p-type doping. Based on this, it is essential to use low WF electrodes and suitable donor materials for efficient electron injection. One of the most popular interfaces for electron injection in conventional OLEDs is organic/alkali metal/aluminium; however, the electronic structure at the interface is complicated due to intercalation of reactive

Fig. 4.8 Schematic diagram of energy level alignment for an ITO/ MoO₃/NPB junction. Reprinted with permission from ref. [22]. Copyright 2009, American Institute of Physics

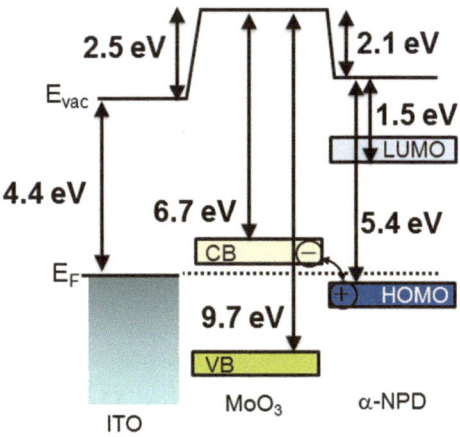

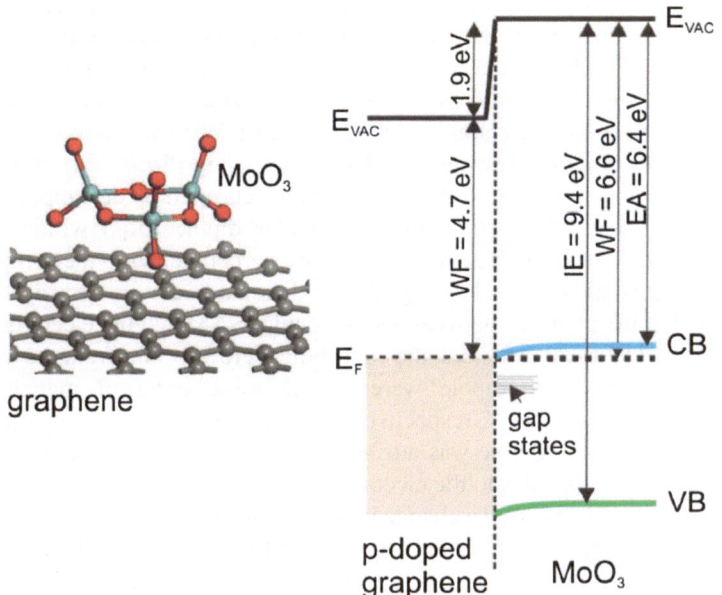

Fig. 4.9 Schematic of a graphene/MoO₃ interface and diagram of energy level alignment for a graphene monolayer/MoO₃ interface. This figure is reprinted from ref. [23]

alkali metals. On the other hand, it is relatively easy to investigate the electronic structure at the electron injection interface in inverted OLEDs since the interface is as simple as that for hole injection in conventional OLEDs. In recent years, the electronic structure at the interface for electron injection in inverted OLEDs has been intensively studied.

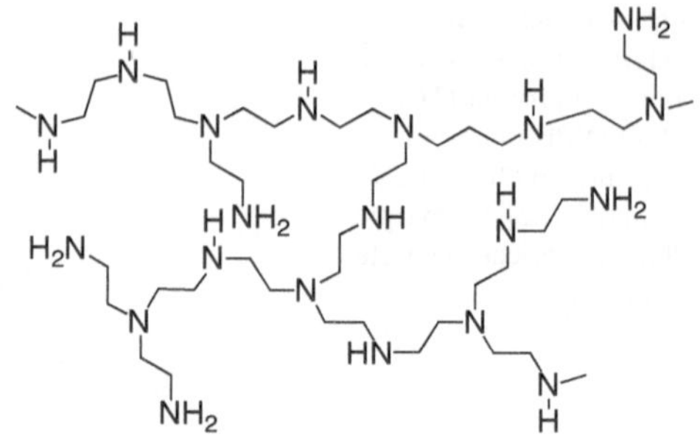

Fig. 4.10 Chemical
structure of PEI

PEI

4.2 Electron Injection Mechanism in Inverted OLEDs

In Chap. 3, electron injection materials suitable for inverted OLEDs are reported. Here the injection mechanisms and related electronic structures will be introduced. As introduced in Chap. 3, amine derivatives such as polyethylenimine ethoxylated (PEIE) and branched polyethylenimine (PEI) have been proposed as interlayer materials, and the related energy level diagrams have been investigated. The first report on the origin of WF shift caused by amine derivatives was published in 2012 [24]. Figure 4.10 shows the chemical structure of PEI. WF reductions of over 1.0 eV on various types of substrates was detected by preparing PEI on various substrates [24]. The mechanism leading to WF shift was also clarified by decomposing it into contributions from (i) the ethylamine molecular dipole (μ_{MD}) within the self-assembled monolayer (SAM) along the direction perpendicular to the surface (leading to an electrostatic potential energy change denoted as ΔV_{MD}) and (ii) the dipole (μ_{ID}) formed at the interface between the molecular SAM and the electrode surface (ΔV_{ID}) (Fig. 4.11). In all instances, the contributions to WF shift from the molecular dipole and the interface dipole were of the same order of magnitude, for instance,–0.5 eV and–0.8 eV, respectively, on Au (111). The contribution to WF shift from the interface dipole was attributed to a slight electron transfer from the amine-containing molecules to the electrode surface.

S. Höfle et al. also reported a change in surface WF by depositing ZnO and a PEI layer on ITO, which was used as the cathode and EIL in inverted OLEDs [25]. Due to the high WF of ITO (4.9 eV), there was a very high barrier of about 1.9 eV for electron injection to the LUMO level of Super Yellow, which is the emitting polymer used in their device. The WF of the cathode was reduced to 4.1 eV by applying a ZnO buffer layer, which resulted in improved electron injection. Actually, the device onset voltage was reduced compared with the device without ZnO. In addition, by combining the ZnO buffer and spacer layer with the PEI interface modifier, the WF of the ITO/ZnO/PEI electrode reached 3.4 eV (Fig. 4.12). The device performance and the onset voltage were drastically improved by matching the bandgap energy of Super Yellow and implying an almost barrier-free electron injection into the emitting layer. Similar electronic structures were reported by Kim et al.

Fig. 4.11 Proposed model for molecular dipole-induced and surface dipole-induced WF reductions on ZnO surface

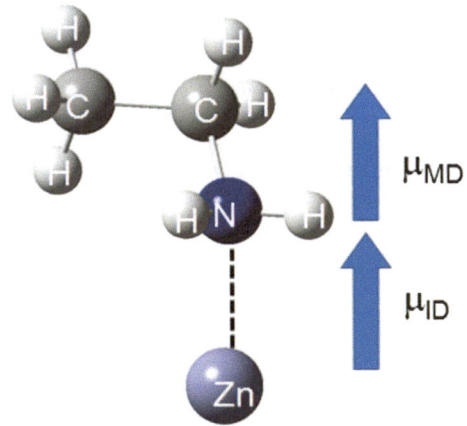

Fig. 4.12 Work functions of ITO, ITO/ZnO, ITO/PEI and ITO/ZnO/PEI electrodes determined by XPS measurements. Reproduced with permission from ref. [25]. Copyright 2014, Wiley-VCH

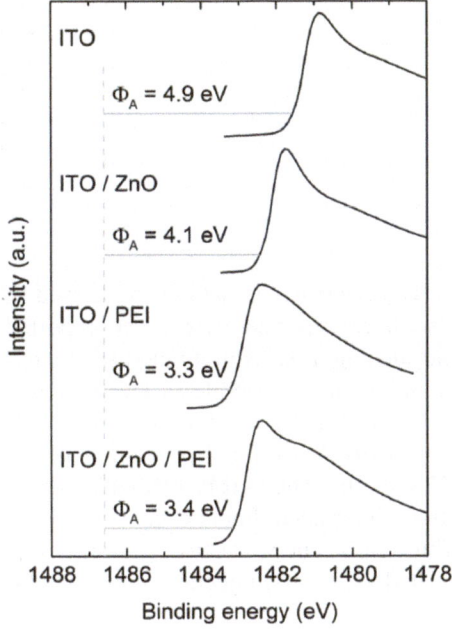

using not only PEI but also PEIE [26]. As the thicknesses of PEI and PEIE increased, the WF of an ITO/ZnO surface gradually increased from 2.47 eV (4 nm PEI) to 3.39 eV (16 nm PEI) and from 3.29 eV (4 nm PEIE) to 3.6 eV (16 nm PEIE) (Fig. 4.13). It was demonstrated that PEI reduced the WF of the ITO/ZnO surface more than PEIE at every interlayer thickness, which is consistent with the UPS results reported by Zhou et al. [24].

Fig. 4.13 (**a**) UPS spectra
for ZnO, ZnO/Cs₂CO₃ and
ZnO/PEI. (**b**) UPS spectra
for ZnO, ZnO/Cs₂CO₃ and
ZnO/PEIE. Reproduced
with permission from ref.
[26]. Copyright 2014,
Wiley-VCH

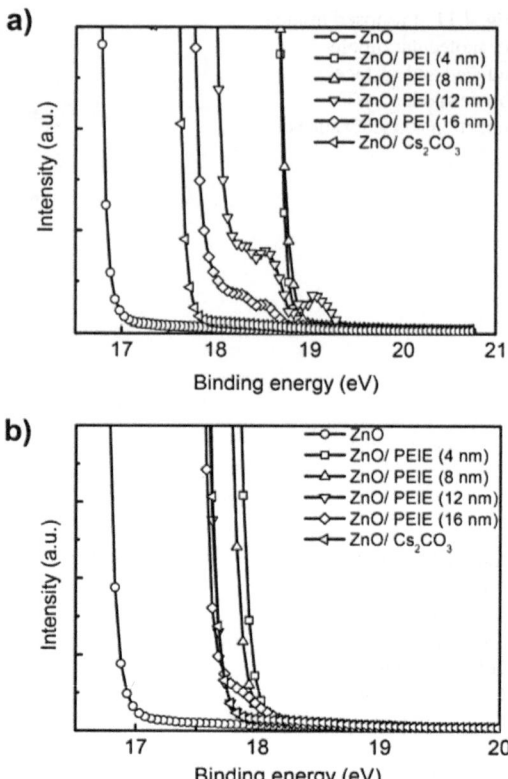

The electronic structures in inverted OLEDs using other amine derivatives have also been reported. Lee et al. reported an interfacial engineering method by introducing a series of amine-based interfacial molecules (AIMs) that contain 2–6 amine groups (2–6 N) [27]. The reduction of the electron injection barrier between ZnO and SY by the formation of AIMs was confirmed by UPS measurements (Fig. 4.14). As the number of amine groups in the AIMs was increased from 2–6, the WFs of the ZnO/AIMs gradually decreased from 4.08–3.78 eV because of the strong interfacial dipole caused by the large number of amine groups in the AIMs. Thus, UPS is now used as a powerful tool for investigating the electronic structures related to electron injection in inverted OLEDs.

4.3   Hole Injection Mechanism in Inverted OLEDs

As introduced in Chap. 2, the first device was constructed by hole injection using molybdenum oxide (MoO₃). The evaluation of hole-only devices revealed the unprecedentedly and asymmetric high hole injection ability of MoO₃. As a result, MoO₃ has been used in many devices and has become a mainstream topic in discus-

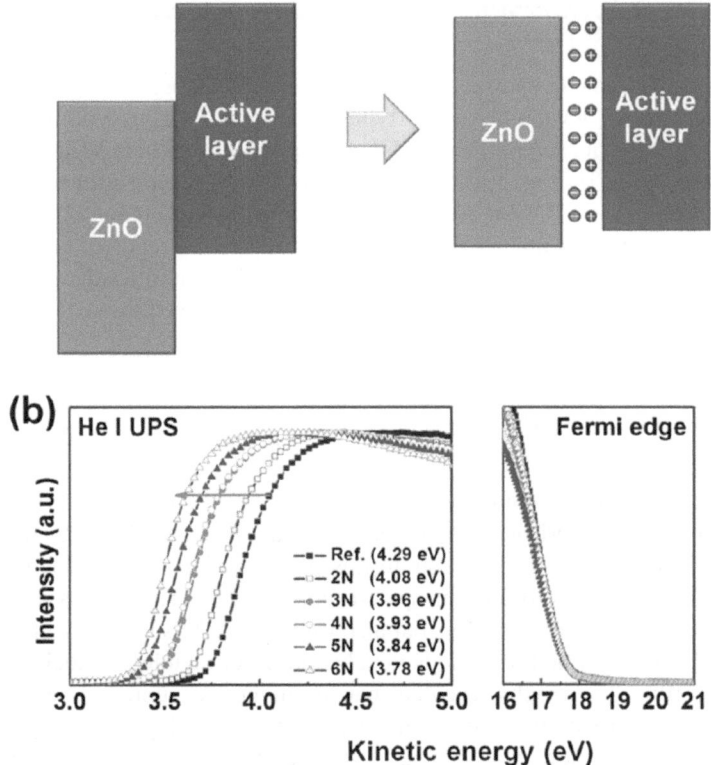

Fig. 4.14 (**a**) Energy diagrams for flat-band conditions at the ZnO/active layer interface with (right) and without (left) AIMs. (**b**) UPS measurements. Reproduced with permission from ref. [27]. Copyright 2015, Wiley-VCH

sions on iOLEDs. In the conventional structure, tungsten oxide [28, 29] and vanadium oxide [20] are reported as other metal oxide materials. Unfortunately, there are few reports of inverted structures, but in principle similar effects can be expected with these metal oxides. The enhancement of hole injection by insertion of a metal oxide enables thinning of the top electrode and also contributed to extract the generated light.

In this subsection, a key point of the high hole injection ability, that is, the difference between organic-on-inorganic and inorganic-on-organic structures, is discussed. Subsequently, poly(9,9-dioctylfluorene-alt-benzothiadiazole) (F8BT), which has been practically applied to hybrid organic–inorganic light-emitting diodes (HOILEDs) and iOLEDs, is introduced in terms of its chemical and electron states on the basis of device characteristics.

4.3.1   Organic/Inorganic Layer and Inorganic/Organic Layer

An organic electronic device has an interface between an organic layer and an inorganic layer. These devices have two types of structure: organic-on-inorganic and inorganic-on-organic. A report showed that the properties of these structures depend on the fabrication process [30]. Inverting the order of deposition (inorganic on organic) induces the diffusion of inorganic materials, which changes both the chemical and electronic structures. In addition, Matsushita et al. reported that the magnitude of the charge transfer of certain types of naphthyl-substituted diamine derivative (NPD) and MoO_3 depends on the molecular orientation [31].

4.3.2 MoO_3 on F8BT

When an energetically symmetric hole-only device [indium tin oxide (ITO)/MoO_3/F8BT/MoO_3/Au] is fabricated, asymmetric conduction characteristics are obtained, as shown in Fig. 4.15. To demonstrate this, the chemical state of each layer was analyzed by X-ray photoelectron spectroscopy (XPS). The results showed that MoO_3 on F8BT has not only Mo^{6+} but also reduced Mo^{5+}, suggesting charge transfer. The electronic structure was also evaluated by ultraviolet photoelectron spectroscopy (UPS). The results revealed that the highest occupied molecular orbital (HOMO) edge shifted owing to the deposition of a MoO_3 layer with a thickness of ~0.5 nm and that a new peak appeared at around 0.5 nm as shown in Fig. 4.16. Figure 4.17 shows the dependence of the intensity of the new peak, which is considered to be due to the new energy level, on the thickness of the deposited MoO_3 layer. The peak intensity is maximum at a thickness of ~2 nm. Moreover, carrier transport properties were evaluated as an example of properties originating from the chemical

Fig. 4.15 Current voltage – voltage curves of hole-only devices

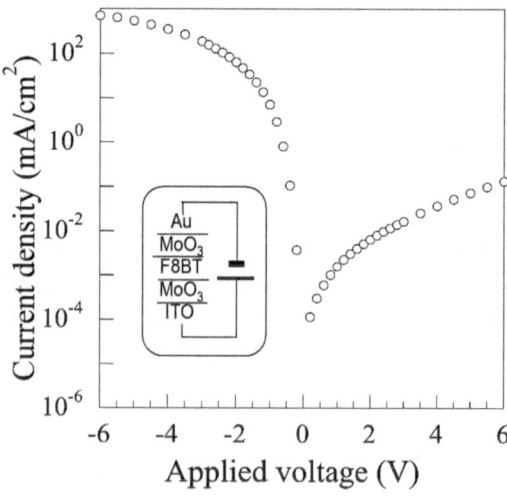

Fig. 4.16 UPS spectra at the valence region of MoO3 deposited on F8BT

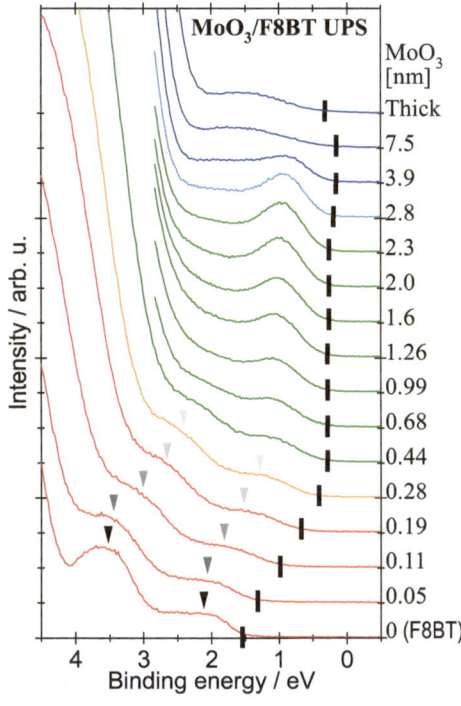

Fig. 4.17 Relative intensity of the new peak in the MoO3 deposition thickness

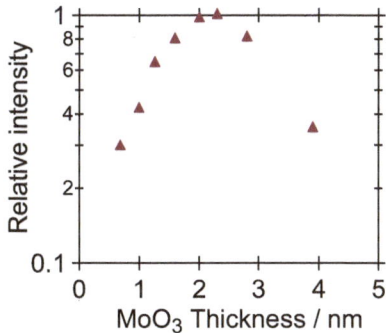

and electronic structures. Gold was deposited as an electrode with a 50 μm gap on spin-coated F8BT as shown in Fig. 4.18. MoO_3 was then deposited on the prepared film. Current (I)-voltage (V) characteristics were measured for each thickness of the deposited MoO_3 layer. The dependence of the I-V characteristics on the MoO_3 thickness is shown in Fig. 4.19. The current did not increase up to a thickness of 3 nm on the glass without F8BT, whereas it increased as a result of the deposition of 0.5-nm-thick MoO_3 on the F8BT. This suggests that electrons were transferred from the organic material to the inorganic material, supporting the results of XPS and

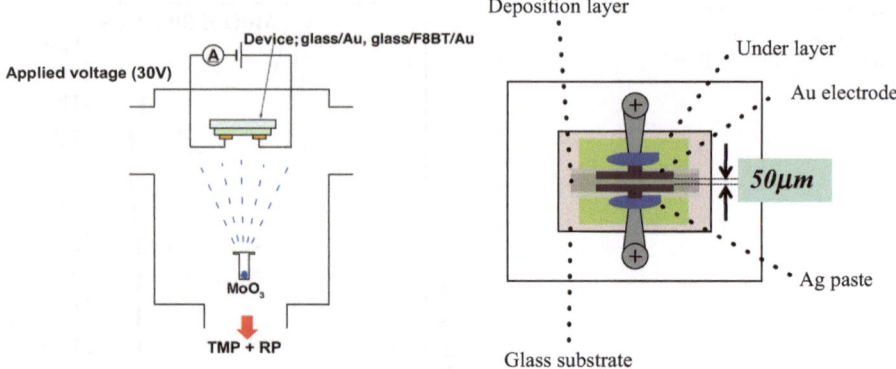

Fig. 4.18 Arrangement of MoO3 deposition details (left) and electrode (right) on the F8BT film

Fig. 4.19 Thickness
dependence of the current
at 30 V in the MoO3
deposition on the F8BT

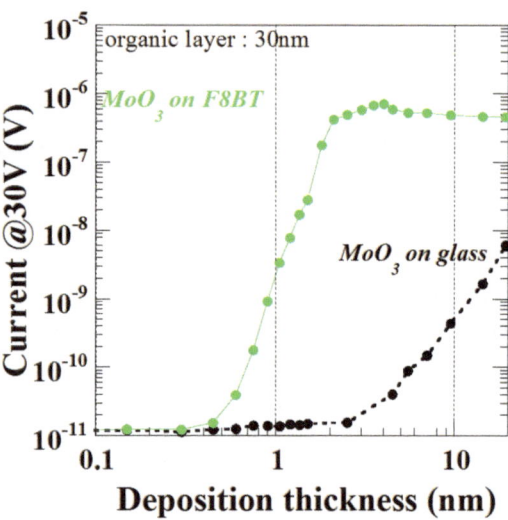

UPS. Once the current started rising, it sharply increased until it saturated. The increase is not monotonic. The measurement indicated that there is an anomaly in the increase in current, which was observed for the deposited MoO_3 layer with a thickness of ~2 nm. Interestingly, this is the thickness at which the maximum intensity of the new peak was observed in UPS. Inorganic-on-organic structures, such as MoO_3 on F8BT, generate major carriers and have a reduced injection barrier, where the magnitude of the changes depends on the physical properties of the inorganic and organic materials and also the fabrication process, although the detailed mechanisms behind these phenomena remain unclear.

4.4 Conclusion

In this chapter, both hole and electron injection mechanisms, as well as the related electronic structures, are introduced. In the case of hole injection, it is essential to make the surface WF of the anode larger. Thus, acceptor materials such as F4-TCNQ, HAT-CN and MoO_3 have been used as HIL, and several photoemission studies have clarified that such acceptor materials can enlarge the surface WF of the anode. In the case of electron injection, on the other hand, it is essential to make the surface WF of the cathode smaller. Thus, metal oxides with low WF and donor materials such as amine derivatives have been used as EIL. Although it was difficult to examine the electronic structures related to electron injection in conventional OLEDs, the electronic structures related to electron injection have begun to be clarified by examining inverted OLEDs. I believe that such understanding of electronic structures related to carrier injection is essential to accelerate the development of carrier injection layers.

References

1. Parthasarathy, G., Shen, C., Kahn, A. & Forrest, S. R. Lithium doping of semiconducting organic charge transport materials. *J. Appl. Phys.* **89**, 4986–4992, 1359161 (2001).
2. Masenelli, B., Berner, D., Bussac, M. N., Nüesch, F. & Zuppiroli, L. Simulation of charge injection enhancements in organic light-emitting diodes. *Appl. Phys. Lett.* **79**, 4438–4440, (2001).
3. Ding, H. & Gao, Y. Au/LiF/tris(8-hydroxyquinoline) aluminum interfaces. *Appl. Phys. Lett.* **91**, 172107 (2007).
4. Ishii, H., Sugiyama, K., Ito, E. & Seki, K. Energy Level Alignment and Interfacial Electronic Structures at Organic/Metal and Organic/Organic Interfaces. *Adv. Mater.* **11**, 605–625 (1999).
5. Mason, M. G. et al. Characterization of treated indium–tin–oxide surfaces used in electroluminescent devices. *J. Appl. Phys.* **86**, 1688–1692 (1999).
6. Lee, S. T., Wang, Y. M., Hou, X. Y. & Tang, C. W. Interfacial electronic structures in an organic light-emitting diode. *Appl. Phys. Lett.* **74**, 670–672 (1999).
7. Pfeiffer, M., Beyer, A., Fritz, T. & Leo, K. Controlled doping of phthalocyanine layers by cosublimation with acceptor molecules: A systematic Seebeck and conductivity study. *Appl. Phys. Lett.* **73**, 3202–3204 (1998).
8. Zhou, X. et al. Very-low-operating-voltage organic light-emitting diodes using a p-doped amorphous hole injection layer. *Appl. Phys. Lett.* **78**, 410–412 (2001).
9. Gao, W. & Kahn, A. Controlled p-doping of zinc phthalocyanine by coevaporation with tetrafluorotetracyanoquinodimethane: A direct and inverse photoemission study. *Appl. Phys. Lett.* **79**, 4040–4042 (2001).
10. Gao, W. & Kahn, A. Controlled p doping of the hole-transport molecular material N,N′-diphenyl-N,N′-bis(1-naphthyl)-1,1′-biphenyl-4,4′-diamine with tetrafluorotetracyanoquinodimethane. *J. Appl. Phys.* **94**, 359–366 (2003).
11. Walzer, K., Maennig, B., Pfeiffer, M. & Leo, K. Highly efficient organic devices based on electrically doped transport layers. *Chem. Rev.* **107**, 1233–1271 (2007).
12. Lüssem, B., Riede, M. & Leo, K. Doping of organic semiconductors. *Phys. Stat. Sol.(a)* **210**, 9–43 (2013).
13. Koch, N., Duhm, S., Rabe, J. P., Vollmer, A. & Johnson, R. L. Optimized hole injection with strong electron acceptors at organic-metal interfaces. *Phys. Rev. Lett.* **95**, 237601 (2005).

14. Witte, G., Lukas, S., Bagus, P. S. & Wöll, C. Vacuum level alignment at organic/metal junctions: "Cushion" effect and the interface dipole. *Appl. Phys. Lett.* **87**, 263502 (2005).
15. Rana, O. et al. Modification of metal-organic interface using F4-TCNQ for enhanced hole injection properties in optoelectronic devices. *Phys. Stat. Sol.(a)* **209**, 2539–2545 (2012).
16. Qi, D. et al. Surface transfer doping of diamond (100) by tetrafluoro-tetracyanoquinodimethane. *J. Am. Chem. Soc.* **129**, 8084–8085 (2007).
17. Lussem, B. et al. Doped organic transistors operating in the inversion and depletion regime. *Nat Commun* **4**, 2775 (2013).
18. Kim, Y.-K., Won Kim, J. & Park, Y. Energy level alignment at a charge generation interface between 4,4'-bis(N-phenyl-1-naphthylamino)biphenyl and 1,4,5,8,9,11-hexaazatriphenylene-hexacarbonitrile. *Appl. Phys. Lett.* **94**, 063305 (2009).
19. Gao, C.-H. et al. Comparative studies on the inorganic and organic p-type dopants in organic light-emitting diodes with enhanced hole injection. *Appl. Phys. Lett.* **102**, 153301 (2013).
20. Tokito, S., Noda, K. & Taga, Y. Metal oxides as a hole-injecting layer for an organic electroluminescent device. *J. Phys. D: Appl. Phys.* **29**, 2750–2753 (1996).
21. Matsushima, T., Kinoshita, Y. & Murata, H. Formation of Ohmic hole injection by inserting an ultrathin layer of molybdenum trioxide between indium tin oxide and organic hole-transporting layers. *Appl. Phys. Lett.* **91**, 253504 (2007).
22. Kröger, M. et al. Role of the deep-lying electronic states of MoO3 in the enhancement of hole-injection in organic thin films. *Appl. Phys. Lett.* **95**, 123301 (2009).
23. Meyer, J. et al. Metal oxide induced charge transfer doping and band alignment of graphene electrodes for efficient organic light emitting diodes. *Sci. Rep.* **4**, 5380 (2014).
24. Zhou, Y. et al. A universal method to produce low-work function electrodes for organic electronics. *Science* **336**, 327–332 (2012).
25. Hofle, S., Schienle, A., Bruns, M., Lemmer, U. & Colsmann, A. Enhanced electron injection into inverted polymer light-emitting diodes by combined solution-processed zinc oxide/poly-ethylenimine interlayers. *Adv. Mater.* **26**, 2750–2754, 2618 (2014).
26. Kim, Y.-H. et al. Polyethylene imine as an ideal interlayer for highly efficient inverted polymer light-emitting diodes. *Adv. Funct. Mater.* **24**, 3803–3814 (2014).
27. Lee, B. R.et al. Amine-Based Interfacial Molecules for Inverted Polymer-Based Optoelectronic Devices. *Adv. Mater.* **27**, 3553–3559 (2015).
28. Li, J. et al. Enhanced performance of organic light emitting device by insertion of conducting/insulating WO$_3$ anodic buffer layer. *Synth. Met.* **151**, 141–146 (2005).
29. Meyer, J. et al. Highly effcient simplified organic light emitting diodes. *Appl. Phys. Lett.* **91**, 113506 (2007).
30. White, T. R. et al. Interface structure of MoO$_3$ on organic semiconductors. *Sci. Rep.* **6**, 21109 (2016).
31. Matsushima, T. et al. Charge trandfer-induced horizontal orientation of organic molecules near transition metal oxide surfaces. *Org. Electro.* **14**, 1149–1156 (2013).